PRACTICAL STUDIES OF
ANIMAL DEVELOPMENT

Practical Studies of Animal Development

F. S. BILLETT
and
A. E. WILD

*Department of Biology,
University of Southampton*

CHAPMAN AND HALL   London

First published 1975
by Chapman and Hall Ltd
11 New Fetter Lane, London EC4P 4EE

© 1975 F. S. Billett and A. E. Wild

Printed in Great Britain by
Willmer Brothers Limited, Birkenhead

SBN 412 10360 5

All rights reserved. No part of this book
may be reprinted, or reproduced or utilized in
any form or by any electronic, mechanical or
other means, now known or hereafter invented,
including photocopying and recording, or in
any information storage and retrieval system,
without permission in writing from the
Publisher.

Distributed in the U.S.A.
by Halsted Press, a Division
of John Wiley & Sons, Inc., New York

Library of Congress Catalog Card Number 74–14712

Contents

Preface	*page*	ix
1 General Requirements		1
Animals		1
Legal considerations		2
The laboratory		3
Equipment		4
Glassware		5
Instruments		6
Sources of information		12
References		12
2 Echinoderms and Ascidians		14
Echinoderms		15
Induced spawning and artificial fertilization		17
Ascidians		20
Ciona intestinalis		22
Dendrodoa grossularia		27
References		28
3 Molluscs, Annelids and Nematodes		29
Molluscs		29
Patella		30
Crepidula		32
Littorina saxatilis		34
Annelids		36
Tubifex		36
Pomatoceros		39
Nematodes		41
Female reproductive system of *Ascaris*		42

	Development of *Rhabditis*	46
	References	48
4	**Insects and Crustacea**	50
	The Locust	54
	Life history	54
	Examination of the ovaries	55
	Development	55
	Calliphora	59
	Maintenance and life cycle	61
	Demonstration of hormonal control of metamorphosis	63
	Drosophila	67
	Life cycle and culture of larvae	67
	Observations of puffing patterns in the polytene chromosomes	68
	Crustacea	74
	Larval forms	75
	Culture of *Artemia salina*	76
	References	77
5	**Fish**	81
	Goldfish	82
	Examination of the gonads	82
	Killifish	83
	General care	85
	Egg collection and rearing	85
	Development of *Aphyosemion scheeli*	86
	Guppies	87
	Artificial fertilization	90
	Experimental Work on Fish Embryos	92
	Decapsulation	93
	Explants of the blastoderm	93
	Experimental analysis of development	94
	References	94
6	**Amphibia**	96
	Xenopus	97
	Induced spawning	98
	Development	101
	Notes on feeding tadpoles	107
	Axolotls	109
	Spawning	110

CONTENTS vii

  Development 112
  Notes on rearing from the larval stage 119
 Experiments on Amphibian Embryos 119
  Artificial fertilization 122
  Mechanical decapsulation 124
  Chemical decapsulation 125
  Exogastrulation 127
  Ectodermal explants from gastrulae 128
  Implantation of the dorsal lip of the blastopore 133
  Optic vesicle transplantation 134
  Other experiments on amphibian embryos 140
 Experiments on amphibian larvae 141
  Lens regeneration from the cornea 142
  Immunoelectrophoretic analysis of lens proteins 149
  Thyroxine induced regression of isolated tadpole tails 159

 References 163

7 Birds 166
 Examination of the early chick blastoderm 167
  Isolation of the blastoderm 168
  Stained preparations 171
 Culture of chick blastoderms 172
 Removal and culture of the blastoderm 173
 Preparation of chorio-allantoic grafts 180
  Isolation of limb buds 181
  Preparation of host embryos 181
 References 183

8 Mammals 185
 The oestrous cycle in the mouse 187
 Preparation of vaginal smears 190
 Examination of unfertilized ova from superovulated mice 193
  Superovulation 194
  Female reproductive system 196
  Isolation of living, unfertilized eggs 197
 Reproductive system of the male mouse 200
  Male reproductive system 201
  Examination of spermatozoa 204
 Examination of mouse embryos up to and including the blastocyst stage 206

Timing of matings	208
Removal of eggs from the oviduct	209
Removal of blastocycts from the uterus	210
In vitro culture of mouse eggs and egg fusion technique	212
Culturing eight cell and earlier stage eggs	215
Removal of the zona pellucida and fusion of eggs	218
Arrangement of the foetal membranes in the rabbit and localization of immunoglobulin in the yolk sac splanchnopleur	219
Examination of the rabbit conceptus	224
Preparation of tissue for treatment with fluorescent labelled antibodies	226
Immunofluorescent staining	227
References	230
Index	233

Preface

The purpose of this book is twofold: it is meant to serve both as a practical manual for the study of animal development and as a general introduction to the subject. Central to our endeavour is the belief that developmental biology is best taught and learnt at the laboratory bench, with specimens which are either alive and can be seen to develop or with fresh material derived directly from the egg (as in birds) or mother (as in mammals). Once the dynamic nature of development is appreciated and the overall structure of the developing organism discerned the more conventional study of sections and whole mounts is more likely to become a delight rather than a difficult, and often meaningless, chore. We have laid considerable stress on the early development of animal embryos and the ways in which they can be obtained from a relatively few, but reliable, sources. In addition, emphasis has been placed on fairly simple experiments which make use of the embryos and larvae chosen for the purpose of illustrating development. Embryology ceased to be a descriptive science at the beginning of this century and any practical course, at whatever level, should attempt to reflect this change. It is true that the analysis of development, particularly the genesis of chordate structure, owed much to the invention of the microtome. However, it is as well to remember that a far greater insight into the nature of the developmental process was achieved by a few simple experiments on the embryos of echinoderms and molluscs, and not by sectioning the embryo from one end to the other.

All the practical instructions given in this book are based on our experience of teaching undergraduates over a period of ten years or more and the material is derived from a core of about 30 different practicals and projects. Nothing is included in the detailed instructions which is not based on our own experience. Our aim has been to include sufficient information about each piece of work to enable students to undertake it and for teachers and technicians to prepare it. We have also tried to provide a more general framework by a relevant introduction to each section and by indicating further experiments of a more advanced kind suitable for more demanding projects and research. Obviously the practicals which we describe vary both in the skills required and in the time needed for their completion, and for these reasons they will need to be matched to the average capability of a student group. Thus, although some of the practicals are suitable for large classes at the level of the sixth form, or for students in their first year of higher education, others should only be attempted by small groups of students and a few are best considered as individual projects.

In a general way the practical studies described in the following pages become progressively more difficult as they proceed from invertebrates to fish and amphibia, and then to birds and mammals. This progression reflects the fact that, by and large, it is far easier to initiate and observe development in certain invertebrates than it is in many vertebrates, and among the vertebrates, the embryos of fish and amphibia are more accessible than are those of birds and mammals. It also reflects a deliberate choice in the design of our practical courses, which has been to use some of the invertebrate material at the beginning of a course to illustrate fertilization and the initial stages of development, before proceeding with observation and experiments on the higher forms. Some major groups of animals have either been omitted completely (e.g. reptiles), or have only received a very limited treatment (e.g. crustacea).

PREFACE xi

The reason for this is either that the animals are difficult to obtain or are not, in our view, especially useful for teaching purposes.

Acknowledgements

We are especially grateful for the helpful and constructive criticism of Professor D. R. Newth who read the manuscript during the course of its preparation. We are also indebted to Dr L. G. E. Bell, Elizabeth Adams and John Baynham for their help in the preparation of the photographs, and to Mrs Mavis Lovell and Mrs Anne Wharmby for their secretarial help. A number of the figures are derived, partly or wholly, from the original work of other authors and we would like to thank the following for their permission to use this previously published material: The Royal Society (Fig. 3.1); Dr Vera Fretter and Professor Alastair Graham, The Ray Society (Fig. 3.2); Gustav Fischer Verlag (Fig. 3.3); Professor V. B. Wigglesworth, Chapman and Hall (Fig. 4.7); Professor J. Green, E. F. and G. Witherby, Wistar Press (Fig. 4.12); Elsevier (Fig. 5.4); W. and G. Foyle Ltd., (Fig. 5.5); Professor P. D. Nieuwkoop and Dr J. Faber, North Holland Publishing Company (Figs. 6.1, 6.2 and 6.3); Professor J. Davies, Josiah Macy Jr. Foundation (Figs. 8.10, 8.11 and 8.12); Dr G. Freeman, Wistar Press (Fig. 4.7). We are particularly grateful to Miss Lesley Moxon for allowing us to use her original drawings as the basis for figs. 4.3, 4.4, 4.5, 6.13 and 7.3. Finally we owe a special debt to those students who have taken part in the Developmental Biology practicals at Southampton and who enabled us to assess the feasibility of much of the work described in this book.

Southampton F.S.B.
March 1974 A.E.W.

1 General Requirements

Animals

Almost all the animals mentioned in this book are easy to obtain and are, in most cases, fairly easy to maintain in laboratories for long periods. In quite a few cases they can be bred in the laboratory without difficulty. This is an obvious advantage, for it is not only convenient to be able to draw on laboratory stocks, but also very desirable not to deplete natural ones whether of the same or a related kind. Thus we feel that it is better to use *Xenopus* than it is to use indigenous newts and frogs, and better to use tropical fish than trout or salmon. Even with marine invertebrates it must not be assumed that there is an inexhaustible supply. For instance, the popularity of *Pomatoceros* for teaching purposes seems to have led to a shortage of these animals in some coastal areas. There is no doubt that more attention will have to be given in the future to the artificial breeding of some of the marine forms which are proving useful for teaching and research. There is no reason why *Pomatoceros* and *Ciona* should not be bred under controlled conditions as easily as oysters and barnacles.

Apart from the availability of the animals another important consideration is the duration and timing of their reproductive cycles. For teaching purposes it is not particularly useful to recommend animals which only breed for a short period during the summer months. The range of material described in this book is such as to make both descriptive and experimental studies of animal development a practical possibility at any time of the year.

Legal Considerations

The use of living animals in the classroom or in the laboratory for observation and simple experiments raises ethical problems which should never be ignored. To some people experiments of any kind, even on the simplest living forms, are unacceptable, and to a few others almost anything can be tolerated in the name of Science, Medicine, or worst of all 'Progress'. There are, however, legal requirements in Great Britain (see Cruelty to Animals Act, 1876) which impose strict limits to experiments on animals. These limits are such as to make any work carried out on animals fall within the normal range of human sensibility. The laws relating to the Cruelty to Animals Act apply only to vertebrates, which strictly speaking should include all members of the subphylum Vertebrata. Practical difficulties arise however, in the case of larval, embryonic and foetal forms. At which stage do they become vertebrates? Unfortunately there is no legal definition, but for interpretation of the Act, the Home Office has adopted that stage at which they become 'free living', i.e. naturally escape from the egg membranes. The Act only applies to experiments which are carried out on living vertebrate animals and which are calculated to cause pain. In these circumstances appropriate Home Office licences can be obtained provided that Section 3 of the Act is observed, i.e. the experiment must be done in order to advance physiological knowledge, or to advance knowledge which will be useful for saving or prolonging life or to alleviate suffering; the experiment must *not* be done solely to acquire manual skill or learn techniques.

Unless a good case can be made out for obtaining licences for students, the interpretation which the Home Office places on vertebrate animals imposes restrictions on a valuable source of material for student practical classes, namely vertebrate larvae. Paradoxically, it allows experiments to be conducted on chick embryos prior to hatching and on mammalian conceptuses removed from the mother, both of which may be considerably more advanced than the vertebrate larva. We

GENERAL REQUIREMENTS 3

have been given to understand that the only procedures described in this book which come under the meaning of the Act, are those related to operations on free living tadpole larvae in order to remove the lens (p. 145). We feel that it is important for any reader of this book who is working in Great Britain, and is in doubt as to the legality of the experiments they wish to perform, to seek the advice of their local Home Office Inspector.

Apart from legal requirements relative to animal experimentation, a proper regard for the welfare of animals of all types should always be shown. Care should be taken to ensure that animals are fed, housed, and handled, in the correct manner and that they are not needlessly subjected to pain or adverse stimuli. Reference to the *UFAW* Handbook (1972) is recommended for anyone in doubt as to how to provide for the welfare of animals commonly kept in the laboratory.

The Laboratory

Successful observations and experiments on living embryos depend as much on suitable laboratory facilities and equipment as they do on a wise choice of animals. The remainder of this introduction is devoted to some general comments and advice on working conditions, apparatus and instruments. In keeping with our emphasis on fresh material, histological and cytological techniques are not dealt with here, and only a few are mentioned in any detail later. These are cases, in connection with particular practicals, in which we have found certain techniques useful either for the preservation of embryos or for making sections of embryos, and they have been described where it is appropriate. Operating procedures and recipes for culture media have been dealt with in a similar way.

Ideally, of course, practical work on embryos is best carried out in a laboratory specially designed for the purpose but as this is usually not possible, the best one can hope for is a laboratory which is equipped for cell biology in general and which does not have to be shared with the very different needs of ecologists and behaviourists. Whatever the circumstances the working areas should be clean, cool and well lit. Above all, it must be stressed that satisfactory work in developmental biology is not possible in many cases if the laboratory temperature is too high. For most work we would recommend a laboratory temperature of between 15 and 18°C. In this connection it is as well to remember that a laboratory crowded with students will warm up considerably over a two or three hour period. Fortunately it is not difficult to cool most laboratories during the autumn or winter simply by switching off the central heating at least 24 hours before a practical is due to start. Summer temperatures often make observations and experiments, especially on marine animals, more difficult, and although shading or opening laboratory windows help, an artificial means of cooling the laboratory is of great advantage.

Successful experimental work on embryos depends a great deal on the ability to develop the skill of manipulating small objects under the low power ($c. \times 10$) of a binocular microscope and to make patient observations over lengthy periods. This type of work is made much more difficult than it really is if insufficient attention is given to the relative heights of seat, bench and microscope. These should be arranged in such a way that a student can sit and work in reasonable comfort for a period of two hours. For most purposes a bench height of about 2 ft 6 in (75 cm) with a stool about 6 in lower, is quite adequate. Adjustable seating is an asset and should be considered for all advanced teaching laboratories.

Equipment

In the laboratory several pieces of equipment, which can be used on a communal basis, are essential for developmental studies. A minimum of two incubators is needed. One of these should be adjustable to run between 12 and 20°C and the other should cover a higher range, between 30 and 40°C. A third incubator covering the 20 to 25°C range is a useful optional extra. It is also important to have a high temperature

GENERAL REQUIREMENTS

oven for sterilizing glassware and instruments, and an autoclave for sterilizing solutions. Manufacturers offer a wide range of suitable equipment and detailed comments are unnecessary, suffice it to say that when it comes to making a choice the experience of a previous user is often the best guide.

Having dealt briefly with general laboratory facilities it is necessary to comment on equipment for individual use, namely, microscopes and the instruments used for handling the material described in succeeding sections.

Much of the work described in this book can only be carried out with the aid of a binocular microscope giving good stereoscopic vision. With this type of microscope particular attention should be paid to the magnification range, the size of the field of view and the method of illumination. For most purposes a useful magnification range is from $\times 10$ to $\times 50$. The lower magnification should accommodate an object of about 2 cm in diameter (e.g. a young chick blastoderm) and at the higher magnification something of the order of 5 mm in diameter (e.g. a *Xenopus* egg with its outer jelly coat) should fill the field of view. For work on small eggs, those of marine invertebrates for instance, magnifications of up to 100 times are useful, but the higher the magnification the less will be the depth of focus and the greater the difficulty of manipulation. Most modern dissecting microscopes possess a 'zoom' lens system. Care should be taken to choose a system which requires the minimum amount of refocussing on changing magnification. Good illumination from above the stage is most important. A light source for this purpose may be built into the microscope but an independent, adjustable source is sometimes an advantage, offering greater flexibility in terms of intensity and position of the source in relation to the stage. Heating of the stage by the light source must be avoided by including a heat filter in the system.

Glassware

Certain items of glassware are frequently used in experimental embryology and require special mention. These are the containers in which embryos are observed, stored and operated on. Commonly used are petri dishes, usually 9 cm or 4 cm in diameter, and large and small solid watch glasses. Apart from these most of the other items likely to be used are standard laboratory equipment, such as beakers, measuring cylinders, graduated pipettes, volumetric flasks and so on. Needless to say all glassware should be clean and will frequently need to be sterilized, especially the containers used for the embryos. As an alternative to glass, plastic can be used for some of the items, for example petri dishes, which can be obtained in presterilized packs. Plastic containers however, can only be used a few times, are usually not very good optically, tend to be expensive, and finally present a problem of disposal.

Instruments
Work with embryos and related material involves the use of instruments designed for small scale operations. Some of these instruments such as small scissors, iridectomy scissors, small scalpels and small dissecting forceps are well catalogued by commercial suppliers and do not require special comment. There are, however, certain instruments which might be regarded as the characteristic 'tools of the trade' of experimental embryology, and these need to be described in some detail.

In addition to those which can be obtained commercially a number of simple instruments will need to be made in the laboratory. These include specially sharpened needles, hair loops and small glass pipettes.

Watchmaker's Forceps
Of the instruments which are manufactured, exceptionally fine forceps known as watchmaker's forceps are essential. They are particularly useful, for instance, for handling embryonic membranes. This kind of forcep is illustrated in Fig. 1.1. The very fine points of these forceps need to be resharpened from time to time. This process should be carried out under a binocular microscope using a very fine grade, oiled stone. Care should be taken to sharpen the points evenly by keeping the

GENERAL REQUIREMENTS

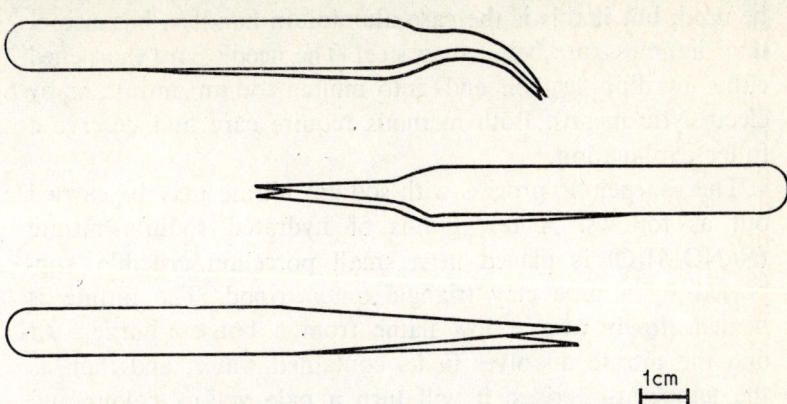

Fig. 1.1. Examples of fine (watchmakers) forceps

points closed during the sharpening process. When not in use the points should be protected by the plastic sheath which is usually provided, or if it is not, by a piece of small bore rubber tubing.

Needles

Both glass and metal needles (Fig. 1.3) can be used for microsurgery. Glass needles may be prepared by drawing out thin (3 to 4 mm diam.) soft glass rod to a very fine diameter. Short lengths of this material are broken off to form needles between 5 and 10 mm long. A suitable length of soft glass rod (about 5 mm diam.) which is tapered to a diameter of 1 to 2 mm at one end, forms the holder of the needle which is attached to the tapered end of the rod using a microburner. Although glass needles can be produced quickly and easily they have obvious disadvantages; they break easily and cannot be sterilized by flaming. Our own preference is for metal needles made from tungsten wire about 0.25 mm in diameter. To make the needles, short lengths of wire, about 2 to 3 cm long, are sealed into glass rods of about 5 mm diameter and about 15 cm long. Soft glass is preferable. The end of the rod is heated until molten and the wire simply pushed into the end of the softened rod to a depth of 3 to 4 mm. Alternatively metal needle holders may be used, but if this is the case aluminium handles, because of their lightness, are better than steel. The needles are sharpened either by dipping their ends into molten sodium nitrate or by electrolytic means. Both methods require care and deserve a fuller explanation.

The sharpening process with sodium nitrate may be carried out as follows. A few grams of hydrated sodium nitrate ($NaNO_3.H_2O$) is placed in a small porcelain crucible, supported by a pipe clay triangle on a tripod. The nitrate is heated *slowly* over a low flame from a bunsen burner. At first the nitrate dissolves in its contained water, and then as the temperature rises, it will turn a pale yellow colour and begin to bubble slightly. At this stage the tip of the tungsten needle should be dipped into the molten nitrate. If the temperature is sufficiently high an exothermic reaction will take place and the wire will suddenly become red hot. After a few seconds the wire should be withdrawn and examined under a binocular microscope to assess the sharpness of the point. If not satisfactory the process should be repeated. Small beadlets of sodium nitrate will adhere to the needle as it cools but these are soluble in water and easily removed before the needle is used. This process is somewhat uncertain; if the temperature is too low hardly anything will happen, and if it is too high the wire will rapidly dissolve. However, the method is a very quick and simple way of producing needles and with a little practice produces excellent results. Some caution is required as the violent nature of the reaction results in a fine spray of molten nitrate in the vicinity of the crucible. We do not consider protective gloves necessary but it is a good idea to wear glasses. Porcelain crucibles will be weakened by the hot nitrate and as soon as they show signs of pitting they should be discarded. This method of sharpening needles is based on that described by Dossel (1958).

A slower but more controlled way of sharpening the tungsten wire is electrolytically. In this method, a pair of unsharpened tungsten needles are mounted in metal holders and each is immersed to a depth of about 5 mm in a small beaker

GENERAL REQUIREMENTS

containing 1N NaOH. The holders are connected to a low voltage d.c. supply. In practice we have used two 45v radio batteries or the low voltage output of a transformer for a microscope lamp. Hydrogen is liberated at the anode and the tip of the wire forming the anode is slowly etched away to form the needle. The rate of etching can be controlled by adjusting the electrical supply and the form of the etching depends on the angle and the distance between the two wires. Like the previous method, this one requires a little practice to produce good results. For building up a stock of good needles the electrolytic method is probably best, but it is a much slower procedure than that using sodium nitrate. The latter is obviously the method of choice for sharpening needles during operations.

Hair loops

The hair loop is a simple device for holding and handling certain embryos during operations. It is made by inserting the two ends of a loop of fine soft hair into the aperture (about 1 mm diam.) of a drawn out pipette and then sealing the hair into its holder by dipping the end of the assembly into liquid paraffin wax (m.p. $c.60°C$). The wax is drawn up into the end of the tube by capillary action and seals the hair into position. The procedure is illustrated in Fig. 1.2. Blond hair taken from a young baby is most suitable for this purpose. Most other types of hair are too springy and tough and will cut through soft embryonic tissues. Loops of about 1 mm are the most useful diameter. Before use, the wax is removed from the loop using a pair of fine forceps. When not in use the loops are protected by storing them in test tubes, a rubber or cork bung being used to support and position the loop in the tube. A similar method of storage can be used for the tungsten needles.

Pipettes

Pipettes of various shapes and sizes are also essential (Fig. 1.3). The simplest are short pieces of glass tubing about 10 to 15 cm long and with internal diameters of about 3 to 5 mm. As in

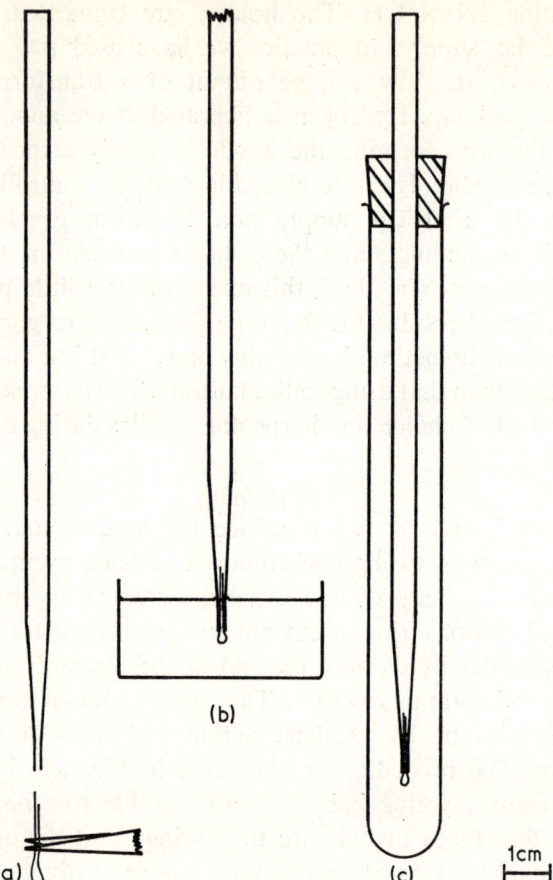

Fig. 1.2. *Making a hair loop* (a) Components; (b) Fixing the loop in position; (c) Completed loop in holder

other cases the freshly cut ends of these tubes should be rounded off in a bunsen flame before use. Pipettes of narrower bore at the working end are made by drawing out standard glass tubing of the dimensions mentioned above. To make these pipettes a fish tail burner is used to soften the centre portion of a suitable length of tubing which at the appropriate time is subjected to a steady straight pull. When cool, the narrowed region of the tube is broken at about 2 to 3 cm from

GENERAL REQUIREMENTS 11

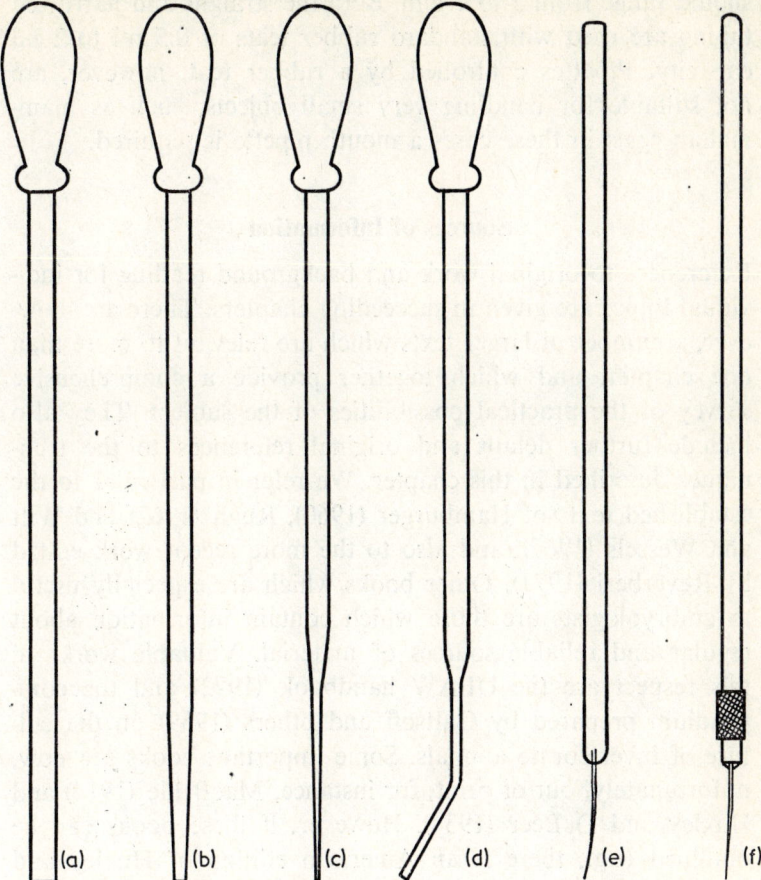

Fig. 1.3. Medium and fine bore pipettes (a–d) and tungsten needles in glass (e) and metal (f) holders

the shoulder of the original tube. For some purposes it is useful to form an angle of about 45° midway along the narrowed portion of the pipette. Hard or soft glass can be used, and although the former is more difficult to work, its melting point being appreciably higher, it usually produces a better result. Whatever the nature of the glass, the original tube must have a fairly heavy wall or else the narrowed piece will be too fragile for effective use. Internal diameters for fine pipettes should range from 2 to 4 mm. Both the straight and narrowed tubing are used with standard rubber teats of 0.5 ml to 2 ml capacity. Pipettes controlled by a rubber teat, however, are not suitable for handling very small objects, such as mammalian eggs; in these cases a mouth pipette is required.

Sources of Information

References to original work and background reading for individual topics are given in succeeding chapters. There are, however, a number of larger texts which are relevant to more than one chapter, and which together provide a comprehensive survey of the practical possibilities of the subject. They also include further details and original references to the techniques described in this chapter. We refer in particular to the established texts of Hamburger (1960), Rugh (1962) and Wilt and Wessels (1967), and also to the more recent work edited by Reverberi (1971). Other books which are especially useful to embryologists are those which contain information about regular and reliable sources of material. Valuable works in this respect are the UFAW handbook (1972) and the compendium prepared by Galtsoff and others (1959) on the culture of invertebrate animals. Some important books are now, unfortunately, out of print, for instance, MacBride (1914) and Huxley and DeBeer (1934). However, if these books can be obtained (e.g., there is an American edition of Huxley and DeBeer, published by Hoffner, Connecticut, 1963), they are worth reading as they not only contain much useful information, but also serve to remind us how much our present understanding of the development of animals is dependent on the practical studies of the past.

REFERENCES

Dossel, W. E., (1958), 'Preparation of tungsten micro-needles for use in embryological research', *Lab. Invest.*, 7, 171–173.
Galtsoff, P. S., Lutz, F. E., Welch, P. S. and Needham, J. G. (1959),

Culture Methods for Invertebrate Animals, Dover Publications, Inc., New York (Unabridged and unaltered republication of the first, 1937, edition).

Hamburger, V. (1960), *A Manual of Experimental Embryology*, University of Chicago Press, Chicago.

Huxley, J. S. and DeBeer, G. R. (1934), *The Elements of Experimental Embryology*, Cambridge University Press, Cambridge.

MacBride, E. W. (1914), *Text Book of Embryology*, Volume I *Invertebrata*, Macmillan, London.

Reverberi, G. (Ed.) (1971), *Experimental Embryology of Marine and Fresh-water Invertebrates*, North Holland Publishing Company, Amsterdam and London.

Rugh, R. (1962), *Experimental Embryology*, Burgess Publishing Company, Minneapolis.

Universities Federation for Animal Welfare (1972), *The UFAW Handbook on the Care and Management of Laboratory Animals*, Churchill Livingstone, Edinburgh.

Wilt, F. H. and Wessells, W. K. (Eds) (1967), *Methods in Developmental Biology*, Crowell, New York.

2 Echinoderms and Ascidians

Although there is no resemblance in form between adult echinoderms and ascidians, we have chosen to link them together in this chapter for three reasons. First it is necessary to emphasize that they both belong to the Deuterostomia. This is a classification of animals which links the echinoderms with the chordates, and is based on the fact that during development, the anus forms at the blastopore (the opening of the archenteron) and the mouth appears as a second, but quite separate, opening. By contrast, development in the other invertebrate phyla, the Protostomia, results in the mouth opening in the region of the blastopore.

The second reason is that it is instructive to compare the early embryology of echinoderms and ascidians. Initially, there is a similar pattern of radial cleavage which contrasts strikingly with the spiral pattern of annelids and molluscs. But the similarity of the form of cleavage in echinoderms and ascidians hides an important difference in egg structure which can be demonstrated experimentally. If the first two blastomeres of the sea urchin are separated, they will, as shown by Driesch in 1900, develop into half-sized, but complete larvae. In contrast, each separated blastomere of an ascidian only produces half a larva (Conklin, 1905). The eggs of ascidians thus lack the capacity of regulation shown by those of echinoderms, and in this respect resemble the eggs of annelids and molluscs. In both ascidians and echinoderms, cleavage leads to a blastula and then to a gastrula. In ascidian embryos these stages are reached when, compared with echinoderms, relatively few cells have been formed: a difference which is possibly related to the difference in egg structure referred to above.

The third reason for dealing with ascidians and echinoderms

together, relates to our practice of using them as reliable material for an introductory study of fertilization and early development.

In this chapter, and the following one, we have concentrated on the descriptive side of development and have made no more than a passing reference to possible experiments with the embryos mentioned. This might be regarded as a serious omission in view of the important contribution that the experimental analysis of echinoderm and ascidian development has made towards our general understanding of the subject, but it is in keeping with our expressed intention in this book not to give details of experiments of which we have had little practical experience. For experimental investigations, such as parthenogenetic activation of eggs, effects of centrifuging, and separation of the blastomeres at the cleavage stage, reference should be made elsewhere (for example, Berg, 1967). For a general appraisal of work on these embryos it is worth consulting Giudice (1973), for echinoderms, and Reverberi (1971), for ascidians.

Echinoderms

One of the simplest and most effective demonstrations of the initial stages of animal development is provided by *Echinus esculentus* (the common sea urchin). If reasonable precautions are taken, the artificial fertilization of sea urchin eggs rarely fails and gives a convincing introduction to the study of animal development. The smaller sea urchin *Psammechinus miliaris* has been less commonly used for this purpose but is just as reliable as the larger urchin and has the additional advantage that it survives better under laboratory conditions. The common starfish *Asterias rubens* is well known for its use in invertebrate practicals but in the UK, at least, it has only received limited application as a source of gametes for artificial fertilization.

Echinus esculentus is the largest of the British sea urchins and is commonly found on rocky shores. It comes inshore during a fairly extended breeding season which lasts from spring to summer and during this time it can be picked up at low tide from steep, sheltered rock faces in the laminaria zone. Alternatively it may be collected in shallow offshore waters by skin divers. A mature urchin measures about 10 cm in diameter around its equator. The colour varies, but the test is usually orange-red and bears white or reddish spines, often tipped with purple. *Psammechinus miliaris* is much smaller, being usually less than 4 cm in diameter. It is olive-green in colour and possesses a close array of purple tipped spines which are sharper than those of *Echinus esculentus.* The distribution of *Psammechinus miliaris* is very similar to that of the larger urchin in that it, too, is found on most rocky shores; it is, however, usually more accessible. Both sea urchins can be confused with other species. In the Channel Islands, for instance, the slightly larger *Echinus acuteus* may be mistaken for *E. esculentus,* and the smaller *Paracentrotus lividus* (found in Ireland, Devon and Cornwall) resembles *Psammechinus miliaris.* Positive identification can be made with an appropriate key but this is not really necessary since all the above species will produce viable eggs and sperm. It is only when the possibility of hybridization needs to be specifically excluded or investigated, that a careful check on the species needs to be made.

    Mature sea urchins can be obtained on suitable coasts during the late spring and early summer months, that is from about the middle of March until the end of June, depending on the locality. The end of the spring term, during the Easter vacation, or at the beginning of the summer term, are by far the best times to attempt artificial fertilization of sea urchin eggs, although successful fertilizations can be achieved as late as July. The best place to do the work is in a marine laboratory at a shore station, since freshly collected sea urchins can then be used. If the work has to be performed away from the sea shore, that is as part of a practical course or as a single demonstration, it is much better to obtain sea urchins directly from those marine stations which supply them. Whether collected directly, or obtained at second hand, the urchins should be stored in a large volume of sea water until required. Storage facilities will obviously vary, but even without marine tanks and circulating sea water, provided the water is aerated, changed frequently, and the temperature is kept below about 15°C, the animals can be kept alive and in good condition for many days. It is wise to keep the sea urchins separated in individual containers, such as large plastic washing-up bowls. The advantage of separating them in this way is that if an urchin spawns it can be readily identified as male or female, also, it will not induce spawning in another animal.

Before attempting to induce spawning by the method described below, it is important to distinguish certain relevant external features. When attached, the urchin will have its oral surface pressed close to the substratum. Protruding from the centre of this lower oral surface will be seen the 'teeth' of Aristotle's lantern. These reflect the basic pentamerous symmetry of the animal, although this is more apparent in the five rows of tube feet and in the five small apertures (the gonopores) situated towards the apex of the more pointed aboral surface. The gonopores, as their name implies, are for the external discharge of eggs and sperm. On the aboral surface can be seen a single, small, rounded structure peppered with minute openings, which is called the madreporite. All these features are illustrated in Fig. 2.1.

Induced spawning and artificial fertilization

Three main procedures have been described to induce spawning of mature sea urchins. These are electrical stimulation, injection of KCl, and the direct exposure of the gonads to sea water. Although the first two methods (for references see Tyler and Tyler, 1966) have the advantage of preserving the sea urchins intact, enabling several possible spawnings from the same animal to be obtained, they lack the simplicity of the more drastic procedure which we prefer because of its reliability. The detailed procedure described below is equally

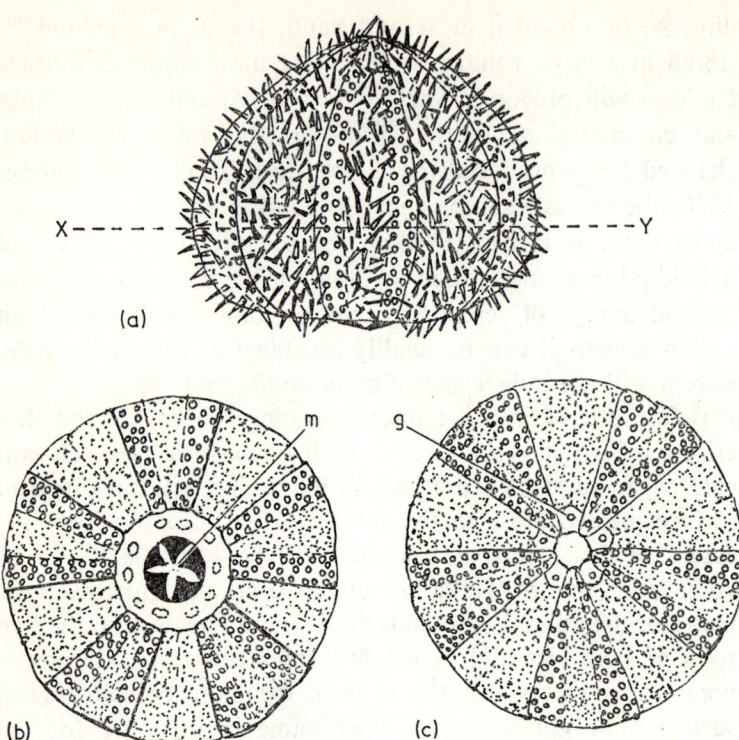

Fig. 2.1. *External features of Echinus* (a) Side view. (b) Oral surface. (c) Aboral surface. X − Y line of cut to expose gonads. *Abbreviations:* m − mouth; g − genital apertures.

suited to *P. miliaris,* taking into account the smaller size of this animal.

The gonads are exposed by cutting through the test of the sea urchin with a small hacksaw blade (about 20 cm long) in such a way as to remove the oral part completely. If the cut is made through the broadest part of the urchin parallel to the oral surface, then the upper two thirds of the animal, containing the gonads, will be obtained. During the cutting operation the animal should be held in a damp cloth. After the oral section has been removed, that part containing the gonads is washed quickly with sea water and supported by the aboral surface, on the rim of a beaker containing sea water. A 500 to

ECHINODERMS AND ASCIDIANS

600 ml beaker is a convenient size for *E. esculentus* and for *P. miliaris* a 50 to 100 ml beaker will suffice. The exposed cavity is filled with clean sea water and if the gonads are ripe, streams of eggs or sperm will start to flow from the genital pores within a few minutes. The method is illustrated in Fig. 2.2. The procedure is repeated until a mature male and female have been obtained. There is then no need to kill further animals, as a single pair will provide sufficient material for several hundred students.

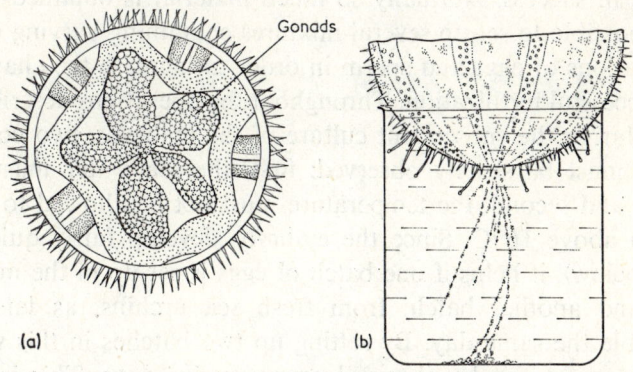

Fig. 2.2 *Method of obtaining eggs or sperm from a sea urchin* (a) Ventral view of urchin after removal of oral half; (b) Aboral half of urchin shedding gametes and supported on rim of beaker containing sea water.

Mature animals have very obvious gonads which appear as five swollen fingerlike bodies occupying the greater part of the exposed cavity. The sexes are easily distinguished; the testes consist of a mass of coiled tubes which are pale yellow in colour, unlike the ovaries which have a fine nodular appearance and are orange coloured. It is instructive to remove small pieces of ovary or testis from freshly killed animals and to examine them under the microscope. This enables an assessment of the state of the gonads to be made and provides an opportunity to look at active sperm and some of the stages of gametogenesis.

The eggs are fertilized in the following way. The beaker containing the eggs is stirred and the egg suspension divided into volumes of about 200 ml. These are placed in large glass finger bowls or similar containers (capacity $\frac{1}{2}$ to 1 litre), 1 to 2 ml of sperm suspension stirred into each volume, and the mixture allowed to stand for ten minutes. It is then diluted with three to four times its volume of sea water. If fertilization has been successful, then the eggs will cleave in about one and a half hours. The above volumes and the suggested dilutions are meant only to serve as a rough guide, but depending on the original concentration of gametes, these proportions usually result in success. Normally so much material is obtained that it is possible to set up several mixtures containing varying concentrations of eggs and sperm in order to increase the chances of successful fertilization. Throughout the operations described, and during the subsequent culture of the embryos, two conditions must be strictly observed: first, the sea water must be clean and second, the temperature must not be allowed to rise much above 16°C. Since the embryos develop fairly quickly (see below), it helps if one batch of eggs is set up in the morning and another batch, from fresh sea urchins, as late as possible the same day. By setting up two batches in this way, development can be observed on successive days. This is important if gastrulation is to be followed. An ordinary light microscope is suitable for observing the developing embryos, particularly if the contrast is increased by partially closing the iris diaphragm of the condenser. Care must be taken to ensure that the light source does not overheat the embryos. It is convenient to keep a batch of embryos in a small petri dish so that they can be sampled occasionally with a fine pipette. Rotation of the container will produce swirling movements and thus concentrate the embryos towards the centre, enabling them to be pipetted up in large numbers. The development of *Echinus* is summarized in Fig. 2.3.

Ascidians

Ascidians are normally less satisfactory than echinoderms for demonstrating the early stages of development. Compared with sea urchins these animals produce only a small number of eggs

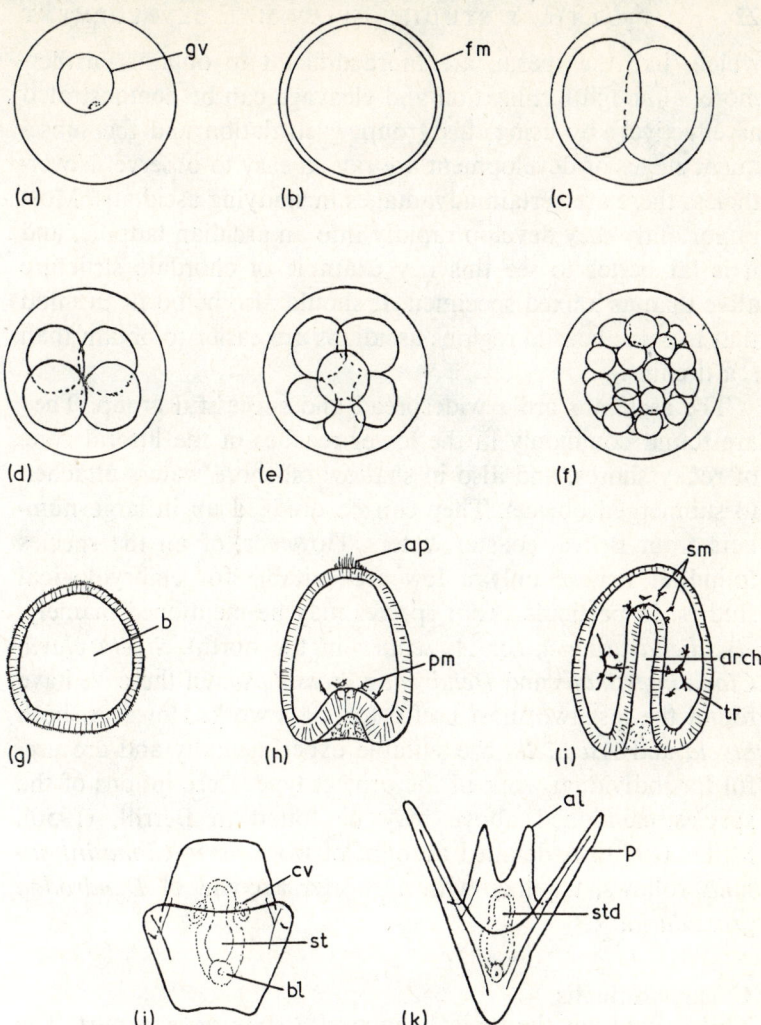

Fig. 2.3. *Stages in the development of Echinus esculentus.* (a) unfertilized egg; (b) fertilized egg; (c) 2 cells (1½ h); (d) 4 cells (2 h); (e) 8 cells (2½ h); (f) 16 – 32 cells (3 – 4 h); (g) late blastula (24 h); (h) beginning of gastrulation (36 h); (i) late gastrula (42 h); (j) early larva (64 h); (k) 4-armed pluteus (5 days). Development at approximately 15°C. *Abbreviations:* al – antero-lateral arm; ap – apical tuft; arch – archenteron; b – blastocoele; bl – blastopore (becomes the anus); cv – coelomic vesicle; fm – fertilization membrane; gv – germinal vesicle; pm – primary mesenchyme; p – post-oral arm; sm – secondary mesenchyme; st – stomach; std – stomodaeum; tr – triradiate calcareous spicule (precursor of skeletal rod).

which, like the sperm, are more difficult to obtain. Furthermore, although fertilization and cleavage can be demonstrated as effectively by using this group, gastrulation and the subsequent stages of development are not so easy to observe. Nevertheless there are certain advantages in studying ascidians. Most importantly they develop rapidly into an ascidian tadpole, and it is far better to see this key example of chordate structure alive than as a fixed specimen. It should also be borne in mind that in some coastal regions ascidians are easier to obtain than sea urchins.

The ascidians are a widespread and successful group. They are found commonly in the lower reaches of the littoral zone of rocky shores, and also in shallow off-shore waters attached to submerged objects. They can be dredged up in large numbers from British coastal waters. However, of all the species found in Britain only a few are suitable for embryological studies. In particular, four species may be mentioned, namely, *Ascidiella aspersa,* (or *A. scabra* in the north), *Styela clava, Ciona intestinalis* and *Dendrodoa grossularia.* Of these we have found the last two most useful for class work. However, both *Styela* and *Ascidiella* are suitable experimentally and are useful for individual work of the project type. Descriptions of the species mentioned above may be found in Berrill, (1950), Millar (1970). A detailed account of work using *Ciona intestinalis* follows, together with a shorter account of *Dendrodoa grossularia.*

Ciona intestinalis
This animal has the typical shape of a solitary sea squirt. The smooth, clean looking, transparent test distinguishes it from other common species, which although of similar size and shape, have a thicker and rougher covering. When fully extended, distinctive orange-red spots may be seen at the openings of both siphons. The animals are usually found in clusters of varying size, and mature specimens range from 4 to 12 cm in length. The species has a wide distribution around British coasts and is found commonly, on submerged and partially

submerged objects, such as wooden posts and buoys. As in the case of the sea urchin, it is best to carry out embryological work on these animals in a laboratory attached to a marine station. However, if this is not possible a direct supply from such a station should be arranged. In an inland laboratory the animals will keep in good condition for many weeks in quite small containers (e.g., domestic washing-up bowls) provided the water is changed fairly frequently, gentle aeration is used, and the temperature kept in the region of 16°C, or below.

A knowledge of the breeding season is essential to ensure that animals containing ripe gonads are obtained. In *Ciona,* as in other ascidians, this appears to be dependent on temperature, breeding animals being found during the summer months in northern waters, and from spring to late autumn in the south-west (Millar, 1952). The condition of the gonads can be determined by a simple dissection, which is also a necessary preliminary to any attempt at artificial fertilization.

The location of the gonads and their ducts need not be a haphazard affair, as the main structures of the animal can be seen through the transparent test if the animal is held up to the light. Before the dissection is made the animal should be laid on its right side in a dissecting dish, i.e. the exhalent siphon, which marks the dorsal side of the animal is on the right (Fig. 2.4). A small round plastic dish about 10 cm in diameter, and containing a layer of wax, makes a useful dissecting dish. During the dissection the exposed organs should be kept moist with sea water, but at no stage should the animal be completely covered. An initial longitudinal cut is made in the test using a pair of fine scissors; it is easier to make this initial cut if the animal is picked up. The incision should extend from the base of the animal to the exhalent siphon. Next, the test is pinned back to expose the thin body wall. Once again the exhalent siphon should lie to the right. A second longitudinal incision made through the body wall will expose the intestinal loop and the branchial basket. If the intestine is not disturbed the sac-like ovary will be seen lying in the loop. This initial

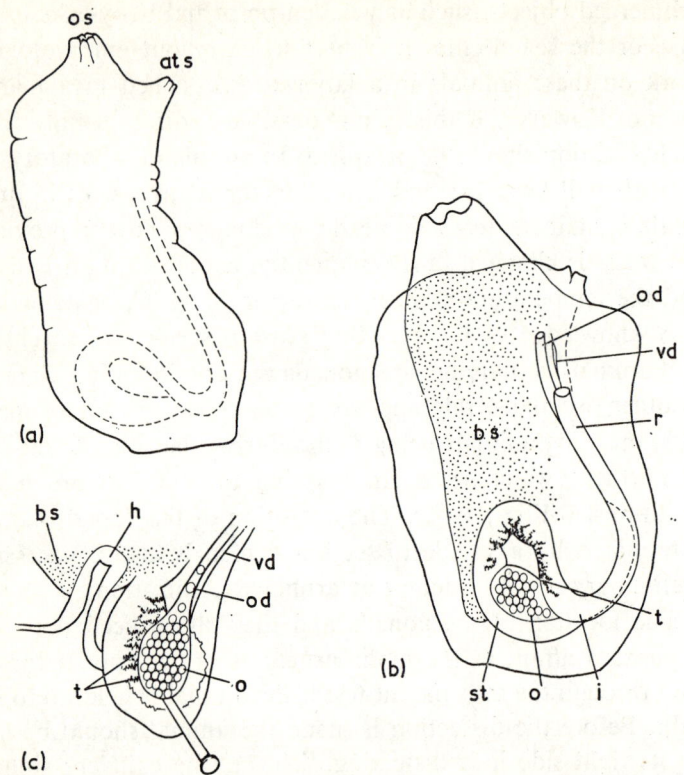

Fig. 2.4. *Ciona intestinalis* (a) External features and indication of position of intestinal loop; (b) Dissection to show gonads and their ducts; (c) Detail of region of gonads. *Abbreviations:* at s – atrial siphon; bs – branchial sac; h – heart; i – intestine; od – oviduct; os – oral siphon; o – ovary; st – stomach; r – rectum; t – testes; vd – vas deferens.

stage of the dissection is completed by deflecting the loop towards the base of the animal and securing it with a pin.

In order to obtain eggs and sperm it is necessary to locate both the oviduct and the vas deferens, which are best seen close to their origin at the gonads. It is not worth wasting time on immature specimens (below about 4 cm long) or on larger animals containing unripe gonads, so it is necessary to determine the condition of the gonads at this stage of the dissection. If the ovary is pigmented it usually contains eggs. Mature testes are seen as a scattered array of milky white vesicles on the wall of the intestine in the vicinity of the ovary. In order to trace the ducts a good hand lens or dissecting microscope, and suitable dissecting instruments, are essential. The oviduct starts as a thin walled tube from the anterior end of the ovary, and later continues alongside the vas deferens, both ducts accompanying the intestine towards the exhalent siphon. The vas deferens is formed by the union of numerous smaller tubes running from the testes. If the gonads are ripe these ducts are very obvious, particularly the vas deferens itself, which will be seen as an opaque white line. The empty female duct is not so obvious as it is thin walled, but the pigmented eggs produced by the ovary are clearly visible through the wall of the oviduct. Each duct should be traced from its origin at the gonads to its opening in the vicinity of the exhalent siphon (Fig. 2.4).

Eggs and sperm are obtained either by squeezing the anterior end of each duct with a blunt pair of forceps, or by cutting the ducts near to their openings. If an exact timing of fertilization is required then it is best to obtain eggs from one animal and sperm from another. It is, however, unnecessary to use eggs and sperm from different animals because if the gametes are mature, self fertilization appears to be as effective as cross fertilization. A pipette with a bore of about 2 mm is used to remove the gametes from the vicinity of their ducts.

To carry out fertilization the following procedure is recommended. Place between ten and twenty eggs into approximately 2 ml of clean sea water contained in a solid watch glass, add one drop of concentrated sperm suspension and mix the gametes thoroughly by gently rotating the container. After five to ten minutes transfer the eggs to another watch glass or small petri dish containing 2 to 5 ml of sea water. Examine the eggs at this stage to assess the progress of fertilization. If this has been successful, the majority of eggs will cleave between one and two hours later if the temperature is about 16°C. After the initial cleavages development is rapid. The ascidian tadpole is usually formed within 24 hours and under good conditions settlement and metamorphosis will occur within two or three days. The development of *Ciona* is summarized in Fig. 2.5.

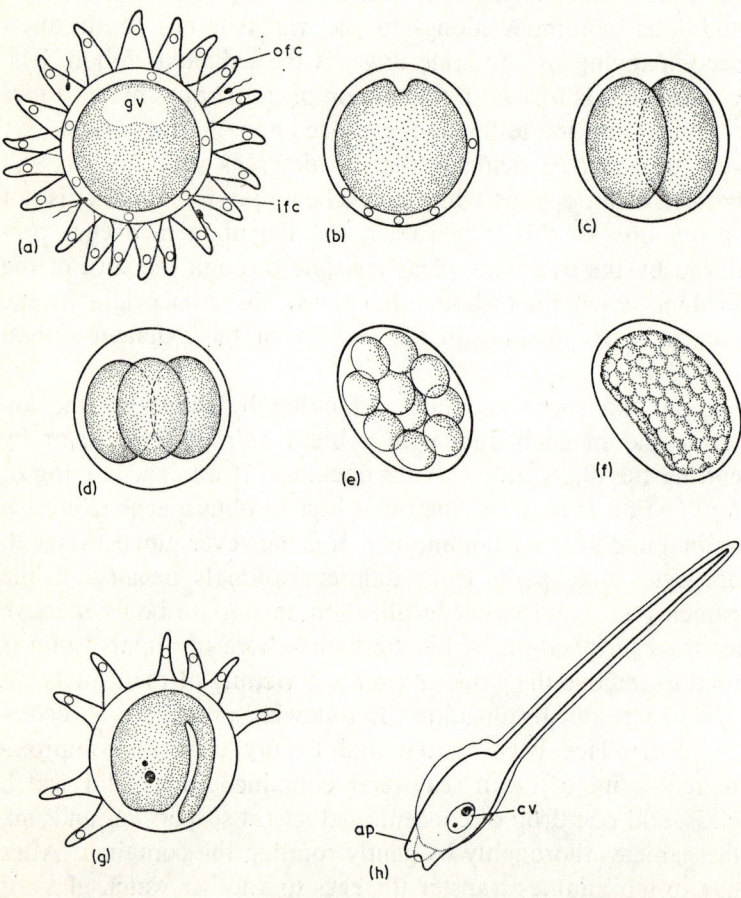

Fig. 2.5. *Stages in the development of Ciona intestinalis* (a) Unfertilized egg; (b) Beginning of the first cleavage; (c) Two cells (2h); (d) Four cells (2.5 h); (e) Early cleavage (3 h); (f) Gastrula (4 h); (g) Stage before hatching (20 h); (h) Ascidian tadpole (24 h); Outer follicle cells are not shown in (b) – (f). Figures in parenthesis are for development at about 16°C. *Abbreviations:* ap – adhesive palps; cv – cerebral vesicle; gv – germinal vesicle; ifc – inner follicle (test) cells; ofc – outer follicle (floatation) cells.

Dendrodoa grossularia

Dendrodoa grossularia (Fig. 2.6) is a small, squat, usually ovoid, ascidian which is found in the lower part of the littoral zone and in the coastal areas of many parts of the British

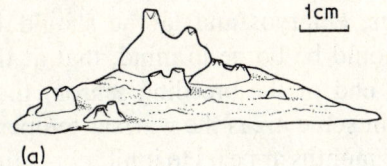

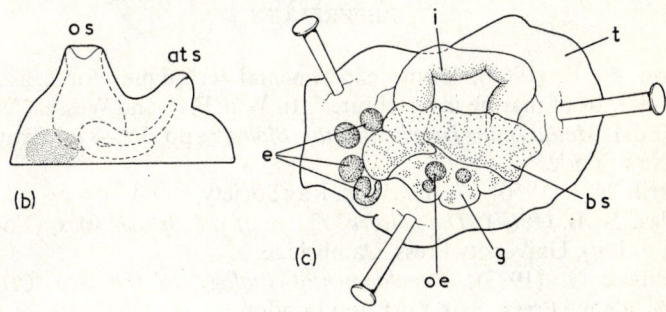

Fig. 2.6. *Dendrodoa grossularia* (a) Group of individuals on shell; (b) View of intact animal from left hand side indicating internal structure; (c) Viewed from beneath with base of animal cut open to show position of eggs and embryos. *Abbreviations:* at s – atrial siphon; bs – branchial sac; g – gonad; e – eggs and embryos; i – intestine; oe – ovarian eggs; os – oral siphon; t – test.

Isles. It frequently occurs as small colonies attached to shells and stones in dredged material. Individuals vary in size from about 2 to 12 mm in diameter and in colour from yellow to red (the usual colour), or brown (Eales, 1967; Millar, 1970).

Being ovo-viviparous *Dendrodoa* is a very useful source of ascidian embryos and larvae. These are located in the atrial cavity from March until October. All stages of development will be found but will of course only occur in mature animals (above c.4 mm in diameter). To obtain embryos and larvae, individual ascidians are pulled away from the rocks and stones to which they are attached. This needs to be done gently to avoid tearing the body of the animal. Then, after placing in a small amount of sea water, the base of the animals is cut open with the aid of fine scissors and forceps in order to locate the atrial cavity (Fig. 2.6). Eggs, embryos and larvae should be found easily. However, it should be borne in mind, that at the beginning, and towards the end of the breeding season, their numbers will be small, and in some areas the warmer temperature of the sea in the summer months appears to inhibit breeding for a time (Millar, 1954).

REFERENCES

Berg, W. E. (1967), 'Some experimental techniques for eggs and embryos of marine invertebrates', In Wilt, F. H. and Wessels, W. K. (Eds) *Methods in Developmental Biology*, pp 766–776. Crowell, New York.

Berrill, N. J. (1950), *The Tunicata*, Ray Society, London.

Eales, N. B. (1967), *The Littoral Fauna of the British Isles*, (Fourth Edition), University Press, Cambridge.

Giudice, G. (1973), *Developmental Biology of the Sea Urchin*, Academic Press, New York and London.

Millar, R. H. (1952), 'Annual growth and reproductive cycles in four Ascidians', *J. Mar. Biol. Ass. U.K.*, **31**, 41–61.

Millar, R. H. (1954), 'The annual growth and reproductive cycle of the Ascidian *Dendrodoa grossularia* (Styellidae)', *J. mar. biol. Ass. U.K.*, **33**, 33–48.

Millar, R. H. (1970), *British Ascidians*, Linnean Society of London, Academic Press, London and New York.

Reverberi, G. (1971), 'Ascidians'. In G. Reverberi (Ed.) *Experimental Embryology of Marine and Fresh Water Invertebrates*, pp 188–214. North Holland Publishing Company.

Tyler, A. and Tyler, B. S. (1966), 'Physiology of fertilization and early development', In Boolootian R. A. (Ed) *Physiology of Echinodermata*, Interscience Publishers, London and New York.

3 Molluscs, Annelids and Nematodes

The close similarity between the development of many annelids and molluscs makes this a good reason for considering them together. The similarity is very evident in many marine species and the types chosen in this chapter reflect this. Thus, if *Patella* and *Pomatoceros* are compared there is found in both types a spiral cleavage initiating a strictly determinative pattern of development which enables a comparable cell lineage to be traced to the common trochophore larval stage. Here, as in many other cases within the two phyla, particular parts of the larva can be shown to be derived from particular blastomeres formed during cleavage, and the loss of a given blastomere at an early stage will lead to an irreparable and defined defect later on. Nematodes are not so obviously related to annelids and molluscs, either in the structure of the animals themselves or in the form of their embryos. Nevertheless, one finds basic features of nematode development, namely a cleavage pattern which can be considered spiral in its form, and a rigidly established cell lineage, which makes it more than a convenience to consider them with annelids and molluscs. Here, as in Chapter 2, we have concentrated on methods of obtaining embryos and observing their development, believing that it is too difficult to carry out meaningful experiments at an undergraduate level on this material, particularly those important experiments which demonstrate the mosaic nature of the eggs.

Molluscs

The value of molluscs as a rich source of material for the experimental analysis of development has been recognized since the pioneering studies of Crampton (1896) on *Ilyanassa*,

Conklin (1897) on *Crepidula* and Wilson (1904) on *Patella* and *Dentalium*. This early work involved detailed observations of cell lineages, established the determinative nature of molluscan development, and focussed attention on the importance of the pole plasms. Subsequent investigations, particularly on *Ilyanassa* and *Limnaea* (for reviews see Collier, 1966; Clement, 1971; Raven, 1966; Hess, 1971) have led to a modification of the strictly mosaic view of molluscan development, drawn attention to the role of the cortex, enlarged our understanding of organogenesis in the group, and more recently, been concerned with the pattern of division sequences which lead to normal development (see Biggelaar and Boon-Niermeijer, 1973).

Although some of the key experiments used in research studies of molluscan development, such as treatment with lithium salts and centrifugation, could form the basis of advanced class work, it is quite obvious that others, for example, the destruction of individual blastomeres, would cause students considerable difficulty. In this section we have, therefore, confined ourselves to a descriptive treatment of three species, namely, *Patella, Crepidula* and *Littorina saxatilis*, which we have found especially useful as an introduction to molluscan embryology. An alternative is to use *Limnaea stagnalis*. These molluscs can be kept easily in fresh water tanks and, provided they are in good condition, will invariably produce eggs if they are transferred to clean water at a slightly higher temperature than normal (say 23°C instead of 18°C). By placing small plastic rafts (petri dish lids will do) in the breeding containers, timed egg layings can be obtained and a series of stages accumulated for teaching purposes. For a detailed description of the development of this species see Raven (1945, 1946).

Patella

Three species of limpet are commonly found on British shores in suitable rocky areas. The most widespread is *Patella vulgata* which tends to inhabit the upper part of the shore. The second species, *P. intermedia* (*depressa*), occurs mainly in the southwest and is usually found from the middle of the shore downwards. The third species, *P. aspera* (*athletica*), has been reported on most coasts and tends to occur lower down the shore. There is, however, a considerable overlap between these species, and the zonation referred to above should serve only as a very rough guide to their distribution. The distinction between the species is not immediately obvious. Diagnostic characters are the colour of the foot, the marginal tentacles, the internal appearance of the shell, and the form of the radula. A good account of these, and other differences, are given in Fretter and Graham (1962). Interestingly, and importantly for practical work on development, these animals have different breeding seasons. *P. vulgata* contains ripe gametes during the winter months (October-December) but the other two species are summer breeders. It is best to work with *P. vulgata* since the breeding season falls at a time convenient for work during the autumn term and this species has the widest distribution; a randomly collected sample of limpets from the upper reaches of a rocky shore will almost certainly contain a majority of *P. vulgata*.

During the breeding season it is a relatively simple matter to obtain eggs and sperm from mature limpets. First, the live animal must be removed from its shell. This is best achieved by inserting a scalpel blade between the mantle and the shell and cutting through the muscles which attach the animal to its shell. If the animal is in breeding condition the gonads will be immediately obvious lying beneath the visceral mass. The sexes are separate. The ripe testis is a pinky-creamish colour and the ripe ovaries are olive green. To obtain eggs and sperm, the gonads are immersed separately in 3 to 5 ml of artificial sea water contained in a solid watch glass. If the gonads do not shed gametes immediately, the surface of the organs should be ruptured gently with a needle or fine forceps. At this stage it is convenient to examine both eggs and sperm with a microscope.

We recommend the following fertilization procedure. Using

a fine pipette place between 20 and 30 eggs into a solid watch glass containing 1 to 2 ml of artificial sea water; the eggs should be concentrated towards the middle of the container. Add 1 to 2 drops of the sperm suspension from the sea water which has been in contact with the testis and mix the eggs and sperm by rotating the watch glass gently. After a few minutes transfer the eggs with a fine pipette into 3 to 5 ml of artificial sea water also contained in a solid watch glass. Check the progress of the fertilization by removing one of the eggs and examining it carefully under the high (dry) objective of the microscope. If active sperm can be seen attached to the jelly coat surrounding the egg then fertilization has probably taken place.

Successfully fertilized eggs show germinal vesicle breakdown within 20 minutes. This is followed by the appearance of the polar bodies after about one hour, and the first cleavage, producing two cells, during the next hour. To obtain a uniform batch of embryos, the normal two cell stages should be transferred to another solid watch glass or small petri dish, containing clean artificial sea water. It is relatively easy to follow the initial cleavage stages (which are typically spiral) but the subsequent stages are not so clear. A trochophore larva will form after about 24 hours and the veliger stage after about 48 hours. Individual embryos can be examined in a drop of sea water on a microscope slide, and protected by a slightly raised coverslip which will avoid squashing the embryo; alternatively a shallow well slide can be used. The development of *Patella* is summarized in Fig. 3.1.

Crepidula
The slipper limpet, *Crepidula fornicata,* is another convenient source of molluscan embryos and larvae. It can be dredged up in southern and eastern areas off the British coast. The animals are associated in clumps of up to 12 individuals and are at first male (the smaller individuals) and then female (the larger individuals). Each animal clings to the one beneath it in such a way that the genital apertures are adjacent. The limpets take

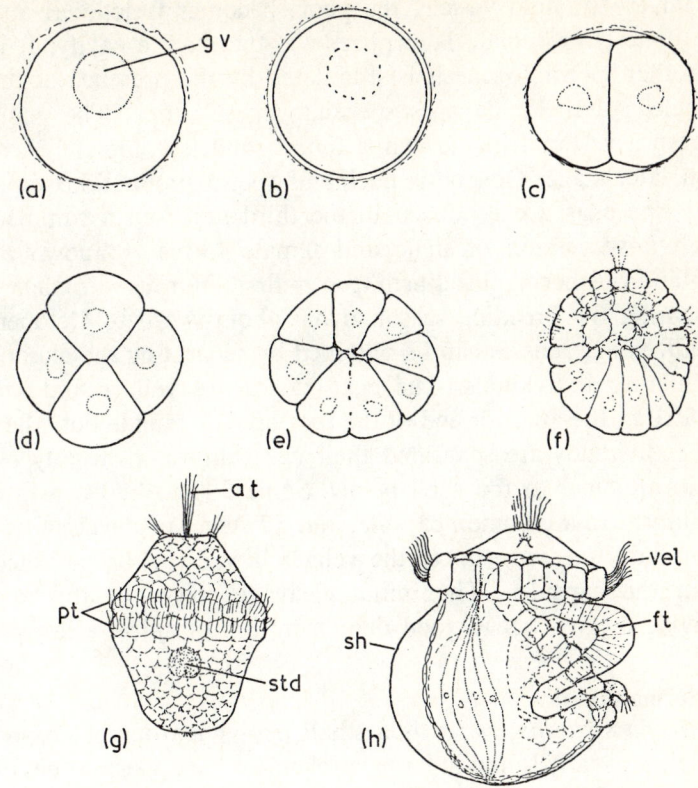

Fig. 3.1. *Stages in the development of Patella* (a) Unfertilized egg; (b) Breakdown of germinal vesicle (20 min); (c) 2 cells (3 h); (d) 4 cells (3.5 h); (e) Early cleavage (6 h); (f) Gastrula (10 h); (g) Trochophore (24 h); (h) Veliger (48 h); Figures in parenthesis are approximate times of development at 15°C. *Abbreviations:* at – apical tuft; ft – foot; gv – germinal vesicle; pt – prototroch; sh – shell; std – stomodaeum; vel – velum. (g and h after Smith, 1935).

4 to 5 years to reach maturity and most individuals live for 8 to 9 years. Only the young limpets move freely and once attached to a group their power of locomotion is lost.

When the young limpet settles its gonad contains both spermatogonia and oogonia. The animal becomes male by further differentiation of the spermatogonia and suppression of oocyte growth. When the gonad produces sufficient sperm to fill the seminal vesicle, the penis becomes functional and the underlying female is fertilized via the mantle cavity. It is said that spermatogenesis is stimulated by the presence of the female. As the male ages spermatogenesis stops. The penis becomes vestigial in the young female and is completely lost at a later stage. Oogenesis begins as sperm production stops, and ripe eggs are produced in the third and fourth summer. Such a succession of male and female forms is known as protandrous hermaphroditism (pro = first, androus = male).

During the breeding season (from about April to October) fertilized egg masses can be obtained by separating the clumps of mature individuals. The egg masses are yellow and are located at the anterior end of the foot of each female but when the individuals are separated their eggs almost invariably remain attached to the shell below. Some 200 to 300 eggs are contained in a common capsule, and all stages of development from early cleavage up to the veliger larvae can be collected from the capsules. The initial cleavage stages of this egg provide a particularly good demonstration of spiral cleavage.

Littorina saxatilis
Littorina saxatilis (*rudis*) is a small ovo-viviparous gastropod which is usually found in large numbers on the upper reaches of rocky shores. It is readily distinguished from similar neighbouring species namely, *L. neritoides* which is much smaller and found higher up the shore, and *L. littoralis* which has a low, obtuse spire, and is found lower down. The nearest to it in shape and appearance is the common edible winkle, *L. littorea,* but this is a larger species and also tends to be found lower down the shore. *L. saxatilis* is distinguished from *L. littorea* not only by its size and ovo-viviparity but also by the angle of the junction between the outer lip of the shell and the main body whorl; in *L. littorea* this angle is acute and in *L. saxatilis* it is rectangular. *L. saxatilis* apparently breeds throughout the year, although in some localities embryos are said to be absent from the brood pouch at certain periods which seem to vary from one locality to another.

MOLLUSCS, ANNELIDS AND NEMATODES

It is easy to obtain embryos at all stages of development from these molluscs. This is achieved by cutting or breaking into the shell, starting at the point where the lip joins the main body of the shell and gradually working back towards the apex; the point of the scissors or forceps must be kept close to the columella. As soon as the first two or three whorls of the shell have been removed the animal itself can be released completely by cutting through the columellar muscle. It is best to attempt to keep the animal intact, so that if it is a pregnant female, the embryos can be located *in situ* in the brood pouch (Fig. 3.2). Finally, the animal is placed in clean sea water and the embryos freed by opening the brood pouch. The whole operation need only take five or ten minutes and will provide a quick and reliable demonstration of molluscan embryology.

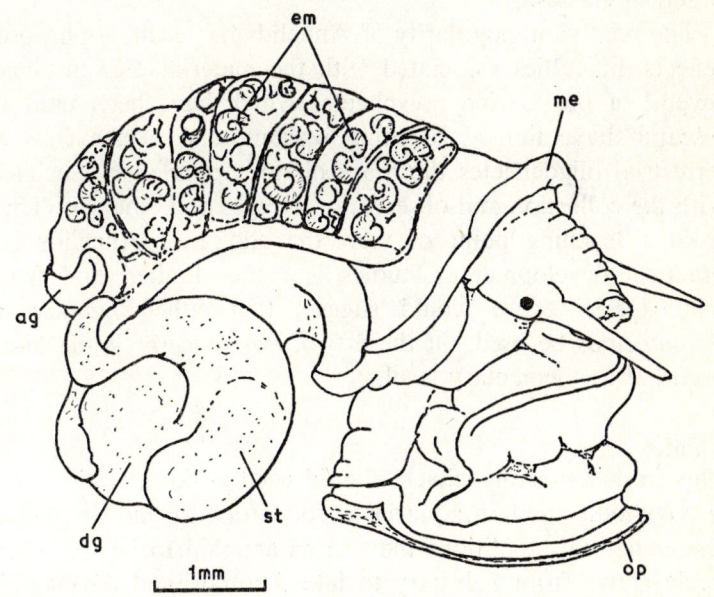

Fig. 3.2. *Littorina saxatilis* Female removed from shell to show location of embryos in brood pouch. *Abbreviations*: ag-albumen gland; em-embryos; dg-digestive gland; me-edge of mantle; op-operculum; st-stomach. (From Fretter and Graham, 1962, British Prosobranch Molluscs, The Ray Society, London).

Annelids

The study of the development of Annelids has made a significant contribution to our understanding of the mosaic nature of the Spiralian egg. In many respects these investigations have been very similar to those carried out on Molluscs (see p. 30). This is especially true of marine forms where the similarity between the methods and results of cell lineage studies are particularly striking. However, in recent times work on Molluscs has overshadowed that on Annelids which have consequently received rather less attention for research and teaching purposes. A recent article by Reverberi (1971) gives a good account of the range of investigations possible with Annelid embryos, but it is quite apparent that the majority of the experiments mentioned cannot be readily adapted for practical classes.

The relative unpopularity of Annelids for teaching purposes reflects difficulties associated with the material. For instance, several of the marine polychaetes which have been used in research have limited breeding seasons, and in the case of terrestrial oligochaetes there are obvious problems associated with the collection and observation of eggs and embryos. Thus from a teaching point of view the choice of Annelids for practical developmental studies is rather limited and from our experience we would suggest that either *Tubifex* or *Pomatoceros* be used. Of these two, *Pomatoceros* is the more useful at an elementary level.

Tubifex

This fresh water oligochaete should need no introduction as it is commonly used in aquaria as food for fish and amphibia. Under natural conditions the worms are said to be reproductively active from February to late Autumn, and during this time (particularly during late spring and early summer) mature individuals, containing swollen ovaries, can be seen. Worms in this condition will produce cocoons readily. These are ovoid (about 1.5 mm diam.) and transparent, and usually contain about 6 eggs. At the right time of the year cocoons will almost invariably be found in clumps of worms which have recently been brought into the laboratory. Once in the laboratory the worms are usually kept in running water and cocoon production can be stimulated by transferring a clump of worms to a beaker containing water which remains unchanged for several hours. Separation of the clumps of worms, by shaking them with forceps, will enable the cocoons to be detected and removed with a fine pipette. The development of *Tubifex* is illustrated in Fig. 3.3. However, it should be emphasized that apart from the earliest stages which are described below, the process is not easy to follow. A detailed account will be found in Stephenson (1930).

Soon after laying, and following fertilization, the eggs will undergo maturation. As a result the first and second polar bodies will be formed and the animal and vegetal pole plasms segregated. This initial stage is associated with the formation of pronounced lobes in the animal region of the egg which indicate changes in the nature of the cortex. These lobes are so pronounced that they can be mistaken for abnormal cleavages; there is no need to discard the eggs at this stage! The first cleavage occurs about 2 hours after laying and divides the egg very unequally into a large (CD) cell and a small (AB) cell. Two hours later the second cleavage begins, the large CD cell divides before the smaller AB cell, giving a 3-cell followed by a 4-cell stage. The animal cells are smaller than the vegetal, and at the 8-cell it is conventional to distinguish a first quartet of micromeres (1a-1d) from the remaining macromeres (1A-1D). The division of 1D gives 9-cells, and the following division of 1C, 10 cells. The formation of the 2nd quartet (2a-2d) is completed by the division of 1A and 1B. The precocious divisions of the CD and the D cells are associated with the presence of the pole plasms which normally become confined to 2d (animal pole plasm) and 2D (vegetal pole plasm). Subsequent divisions confine the animal pole plasm to the precursor cell ($2d^{11}$) of the two ectodermal germ bands and the vegetal pole plasm to the precursor cell (4d) of the two

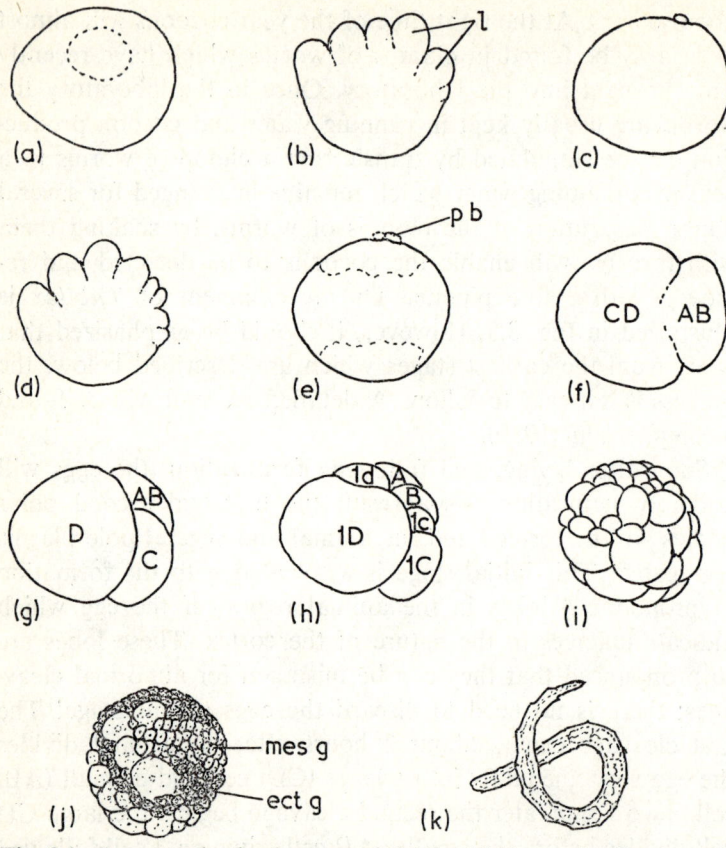

Fig. 3.3. *Stages in the development of Tubifex* (a) Egg at fertilization; (b) Before the formation of the first polar body; (c) First polar body (1 h); (d) Before the formation of the second polar body; (e) Second polar body (1.5 h); (f) First cleavage (2 h); (g) 3 cells (4h); (h) 6 cells (5 h); (i) Late cleavage (10 h); (j) Formation of germ bands (24 h); (k) Young worm; Figures in parenthesis are approximate times for development at about 18°C. *Abbreviations:* ect g – ectodermal germ band; l – lobe like protrusions of egg surface; mes g – mesodermal germ band; p b – polar bodies. (Partly after the figures of Penners, from Stephenson, 1930).

mesodermal germ bands. Clearly the pole plasms play an important role in development. Their importance is confirmed by experiment and is related to the relatively high concentration of cytochrome oxidase and mitochondria in the 2d and 4d cells.

The importance of the D cell and its associated pole plasms in the development of *Tubifex* was established by the experimental work of Penners (1922-1925). Penners demonstrated that if the A, B and C cells were destroyed by u.v. light then D alone could produce a normal, but smaller, worm. Conversely, the destruction of the D cell prevented normal development as the remaining cells could form neither somatoblasts nor mesoblasts. Penners was also able to show that if the eggs of *Tubifex* were heated slightly for a short time then double embryos were formed. Such treatment results in a more or less equal first cleavage, giving two cells each containing pole plasms and thus each capable of producing the germ bands needed to form a complete embryo. In connection with this experiment it is interesting to note that bifid worms are not uncommon in the cocoons of *Tubifex* and that their presence is associated with raised temperatures (Welch, 1921). Such worms remain in the cocoons, trapped by the narrow exit.

Pomatoceros
Pomatoceros triqueter is a common Serpulid polychaete which lives in calcareous tubes on rocks and stones; it is widely distributed around British coasts. This worm is distinguished from others, which also possess calcareous tubes and are equally abundant, by the size and character of its tube. This is larger than that of *Spirorbis, Hydroides,* and *Serpula*, and is strongly keeled, with a characteristic median point at the anterior end. Old tubes with rough anterior ends and without the anterior point are usually empty. If the point is broken off the end of a tube containing a living worm, it will be refashioned during the course of a few days. *Pomatoceros* is said to produce fertile gametes throughout the year and, as it is very easy to induce both sexes to shed their gametes, this animal is an obvious choice for the study of fertilization and early development. Indeed, its popularity (or otherwise) is already assured, as it possesses a niche in a Nuffield Biology practical guide used in sixth forms.

Successful fertilization depends very much on those requirements described for other marine species; that is, the animals must be mature, the sea water free of contaminants and the laboratory kept as cool as possible. To avoid wasting time opening empty tubes an attempt should be made to detect those containing living worms. This is best done by observing a group of tubes under sea water in undisturbed surroundings and waiting until the fans of the worms protrude. However, as previously mentioned, a well fashioned point at the anterior end of a tube is a good indication that it is occupied. The worms are removed from their tubes by carefully breaking open the back end with fine forceps. A blunt seeker is then inserted into the anterior end of the tube and the inhabitant pushed out gently through the broken rear end. After eviction the worms are placed individually into 2 to 3 ml of artificial sea water contained in a solid watch glass. The process is repeated until one worm of each sex, possessing ripe gonads, is obtained. The mature testis is yellowish white in colour and the ripe ovary is dark pink (the colour of the egg pigment). Contact with sea water will liberate both eggs and sperm, which should be examined under the microscope.

Fertilization is achieved by mixing a drop of sperm suspension with about 0.5 ml of egg suspension contained in a solid watch glass. After 5 to 10 minutes the mixture is diluted approximately tenfold with artificial sea water. Successful fertilization is indicated by the breakdown of the germinal vesicle of the egg after 10 to 15 minutes, followed by the appearance of the polar bodies at the animal pole of the egg within the next hour. Each of the first two cleavages takes about one hour, and the embryo should reach the 32 cell stage in about 6 hours (at about 15°C). Beyond the morula stage the development is not easy to follow until the trochophore makes its appearance about 24 hours after fertilization. As indicated by MacBride (1941), and Seagrove (1941), the larvae will develop normally if fed on the diatom *Phaeodactylum,* and eventually metamorphose and settle as young worms. It is thus possible to study the entire life cycle of this worm in the laboratory. Fig. 3.4 summarizes the development of *Pomatoceros.*

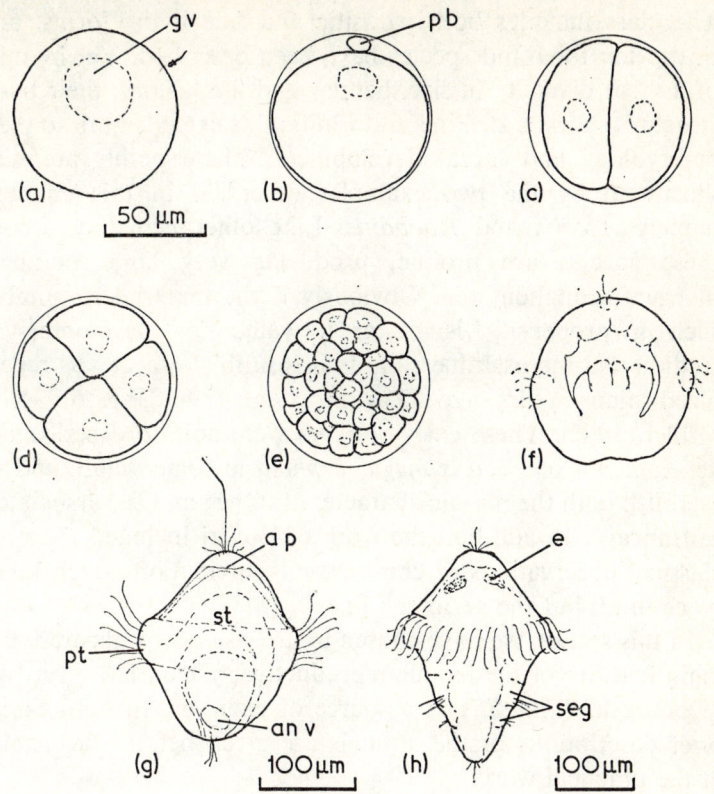

Fig. 3.4. *Stages in the development of Pomatoceros* (a) Unfertilized egg; (b) Formation of the first polar body (1 h); (c) Two cells, viewed from the side (3.5 h); (d) 4 cells, viewed from the animal pole (4.5 h); (e) About 30 cells, viewed from the animal pole (9 h); (f) Gastrula (24 h); (g) Trochophore (2 days); (h) Post-trochophore stage (3 weeks). Figures in parenthesis are for development at about 14°C. *Abbreviations:* ap – apical plate; an v – anal vesicle; e – eye spot; gv – germinal vesicle; pb – polar body; pt – prototroch; seg – initial segment of worm; st – stomach.

Nematodes

The nematodes are a large and important class of animals belonging to the phylum Aschelminthes. The other classes of the phylum contain relatively few species, and apart from the rotifers, little is known about their development. This is in marked contrast to our knowledge of nematode development.

The class includes both parasitic and free living forms, and nearly ten thousand species have been described. Yet in spite of a great diversity in size, habitat and life history, their basic structure shows a striking uniformity; this extends both to their embryology and larval development. These points are well illustrated by the two examples described in this chapter, namely *Ascaris* and *Rhabditis*. Like other nematodes, both these species are prolific, producing very large numbers of eggs throughout life. Obviously if the material is suitable such a propensity is of great value for developmental studies, and the usefulness of *Ascaris* in this respect was recognized many years ago (Zur Strassen, 1896; Boveri, 1899; Müller, 1902). These early workers were able to describe the development of *Ascaris megalocephala* in some detail, and to establish both the mosaic character of its egg and the associated cell lineage. In addition the work of Boveri included the now classical observations of chromosome elimination which takes place in all but the germ cell line.

In this section we have chosen to use *Ascaris* to illustrate the main features of the female reproductive system. However, we recommend *Rhabditis* as a source of embryos. In both cases, brief descriptions of the animals are given before the details of the practical work.

Female reproductive system of Ascaris
The ascarids are large nematodes which inhabit the intestine of vertebrates. Those commonly found in mammals are *Ascaris megalocephala* (=*Parascaris equoruum*), which occurs in horses and cattle, and *Ascaris lumbricoides*, found in pigs and humans. This account concerns *A. lumbricoides* found in pigs (var. *suum*), which although morphologically identical with the species found in man (var. *humanis*), is said to be physiologically distinct. Although there is reassuring evidence that the variety found in the pig will not infect man (Smyth, 1962), this is not certain and care must be taken to avoid accidental ingestion of the eggs when handling the worms.

For the practical, live worms are required and these should

be obtained direct from the local slaughter house. On arrival in the laboratory the material should be well washed in normal saline. Although the worms can be stored in a refrigerator for 24 hours or so, we recommend that they be used as soon as possible. Usually a tangled mass of worms is obtained and it is worth while spending a few minutes sorting the individuals out and measuring their length. Even with relatively few worms it usually becomes apparent that the population can be divided into two size groups, corresponding to females, which are larger, (20 to 35 cm long) and males, which are smaller, (15 to 20 cm long). Apart from their smaller size, the males can also be distinguished from the females by the shape of their rear end which is strongly incurved towards the ventral surface.

In order to dissect the female reproductive system it is necessary to make a longitudinal incision through the dorsal body wall. This means that the worm must be orientated correctly; this is not an easy task as the body is fairly featureless. However, careful examination will reveal the following features. In a fresh worm there is usually a prominent reddish line running down each side of the body. This marks the position of the excretory canals. Fainter lines run along the middle of the dorsal and ventral surfaces, indicating the position of nerve fibres. Both ends of the worm are pointed but three small labial palps (one dorsal and two ventral) surround the mouth and distinguish the anterior end. Once the anterior end has been determined, the surface of the worm should be examined carefully with either a good magnifying glass or a dissecting microscope in order to locate the opening of the combined uteri. Although the opening is minute it can usually be located fairly quickly. This opening lies on the ventral surface of the worm, as does the even smaller excretory pore which is situated a millimetre or two behind the mouth.

Once the correct orientation has been made, the female worm is pinned ventral surface down in a dissecting dish by placing a pin through the body about a centimetre in front of the female opening and another some two-thirds of the way towards the back end of the body. After pinning, the worm is covered with saline. Using fine scissors and forceps the body wall is then opened carefully to avoid cutting the long straight intestine (usually a yellowish green colour) and the coiled white mass of the reproductive system. It is best to start the incision from the front end and gradually work backwards, pinning out the body wall as the dissection proceeds. The paired uteri should be followed from the point where they join to form a short common tube to the outside of the animal, to the region where they narrow and merge with the oviducts. Separation of the coiled mass of oviduct and ovary will reveal that both are a simple extension of the uterus. Basically the female reproductive system is simply a pair of very long tubes, which form eggs at one end and shed them, after fertilization, at the other. Fertilization takes place in the region where the uterus runs into the oviduct. This region of the tube, which is only a few millimetres long, is called the seminal receptacle. Fig. 3.5 illustrates the dissection and the main features of the female reproductive system.

The sequence of events which takes place in the uterus can be determined by removing short lengths of the tube about 5 mm long, beginning near to the junction to the exterior, and continuing to the point where the uterus narrows into the oviduct. The excised pieces should be placed in saline on a microscope slide and squeezed gently to force out their contents. These are then transferred in a drop of saline to another slide. If this is done systematically the region in which the eggs are being fertilized will be obtained and it should be possible to observe the fertilization process itself, including the amoeboid sperm, characteristic of nematodes, and their movement. Although early cleavage stages can sometimes be found in the uteri, development usually takes place outside the body. The eggs will develop *in vitro* if cultured in saline, and it is possible to obtain the initial larval stage under laboratory conditions (Taylor and Baker, 1968). However, for a simple and quick demonstration of nematode development we feel it is better to use *Rhabditis*, an ovo-viviparous species, in

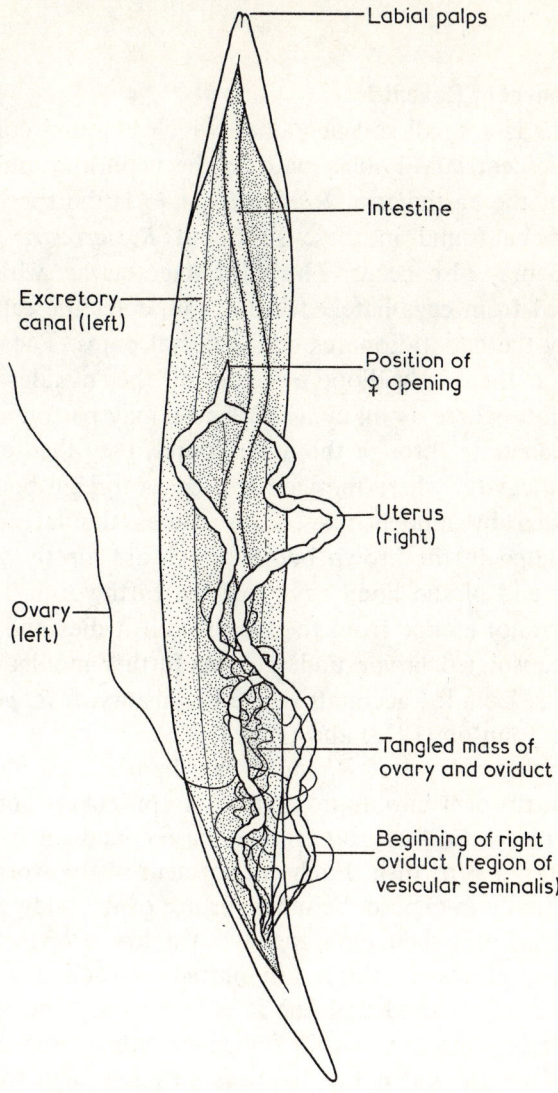

Fig. 3.5. Female *Ascaris* dissected from the dorsal side to show the arrangement of the reproductive system.

which all stages of development can be seen within the uteri of the female.

Development of Rhabditis

Rhabditis is a small soil nematode which is found commonly as a quiescent larval stage both in the nephridia and brown bodies of the earthworm. *Rhabditis pellio* is the species most likely to be found in these sites, but *R. terrestris* and *R. anomola* may also occur. The third stage larvae, which have developed from eggs hatched in the soil, enter the earthworm either by the nephridiopores or the dorsal pores. Those which enter via the nephridiopore settle in the bladder of the nephridium where as many as 12 larvae may be found. If the larvae penetrate through the dorsal pores, they then enter the coelomic cavity, where they are treated as foreign bodies and surrounded by amoebocytes. This causes the larvae to be encapsulated in the brown bodies which are mostly found at the rear end of the body cavity of the earthworm. Thus the larvae cannot escape from the worm until it dies and decays. The encapsulated larvae undergo two further moults to form the adults. Detailed accounts of the life history of *R. pellio* are given by Johnson (1913) and Otter (1933).

A thriving culture of *Rhabditis* can be obtained by cutting up an earthworm into approximately 1 cm lengths and allowing the pieces to decay on an agar base contained in a large (9 cm diam.) petri dish. Each cut segment of the worm is slit longitudinally to expose the inner surface of the body wall and the viscera, and then moistened with a few drops of water. About six pieces of worm are placed in each dish, which should be kept closed and left in a fairly warm place (15 to 20°C). It is difficult to avoid fungal growth on the decaying pieces of worm, and if this happens very few nematodes will appear. For this reason it is best to set up a fairly large number of cultures, and we recommend at least a dozen, each containing about six pieces of worm. In a successful culture young nematodes will appear in four to five days, and within a week large numbers of females containing eggs and embryos should be obtained.

The nematodes are removed from the culture by means of a pipette containing one or two drops of water. In the first instance they should be examined in a cavity slide in order to distinguish between males and females, and to assess the reproductive state of the latter. The sexes are quite distinct, the male possesses a bursa at the rear end, whereas the rear end of the female terminates in a simple point (Fig. 3.6). The females are also larger than the males and possess a distinct genital opening situated half way down the body. Mature females are

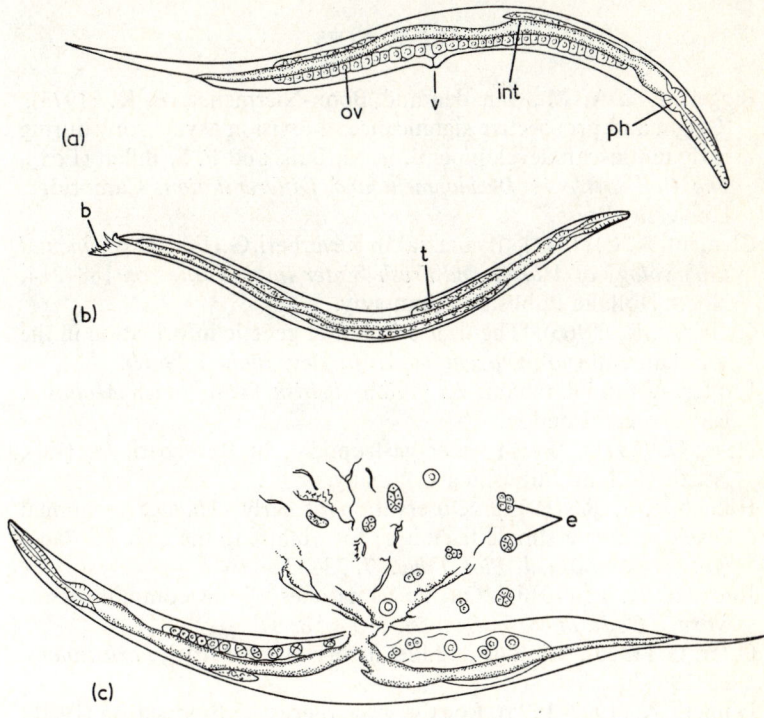

Fig. 3.6. *Rhabditis pellio* (a) Female. (b) Male. (c) Female with released embryos. *Abbreviations:* b – bursa of male; e – embryos; int – intestine; ov – ovary; ph – pharynx; t – testis; v – female opening.

obvious since both uteri contain many eggs and embryos. Once identified, a female containing embryos should be placed in a drop of water on an ordinary microscope slide and a cover slip lowered gently onto the drop. The weight of the cover slip will squash the animal slightly and it should be easy to examine the various stages of development *in situ*. A detailed examination of the embryos can be made by allowing the preparation to dry out. The increasing pressure will release the embryos either via the genital opening or through the ruptured body of the female (Fig. 3.6). A recent account of this technique and further details of *Rhabditis* are given by Hinchliffe (1973).

REFERENCES

Biggelaar, J. A. M., van den and Boon-Niermeijer, E. K. (1973), 'Origin and prospective significance of division asynchrony during early molluscan development', In M. Balls and F. S. Billett (Eds.), *The Cell Cycle in Development and Differentiation*, Cambridge, University Press.

Clement, A. C. (1971), 'Ilyanassa' In Reverberi, G. (Ed.) *Experimental Embryology of Marine and Fresh Water Invertebrates* pp 188–214, North Holland Publishing Company.

Collier, J. R. (1966), 'The transcription of genetic information in the spiralian embryo', *Current Topics in Dev. Biol.*, **1** 39–60.

Fretter, V. and Graham, A. (1962), *British Prosobranch Molluscs*, Ray Society, London.

Hess, O. (1971), 'Fresh water gastropoda', In Reverberi, G. (Ed), North Holland Publishing Company.

Hinchliffe, J. R. (1973), 'Observation of early cleavage in animal development: a simple technique for obtaining the eggs of *Rhabditis* (Nematoda), *J. Biol. Educ.*, **7**, 33–37.

Johnson, G. E. (1913), 'On the nematodes of the common earthworm', *Q. Jl. microsc. Sci.*, **58**, 605–652.

Otter, G. (1933), 'Biology and life history of *Rhabditis*', *Parasitology* **25**, 296–307.

Penners, A. (1922–1925), For these references see Stephenson (1930).

Raven, Chr. P. (1945), 'The development of the egg of *Limnaea, stagnalis* L. from oviposition to the first cleavage'. *Arch. neerl. Zool.* **7**, 91–121.

Raven, Chr. P. (1946), 'The development of the egg of *Limnaea stagnalis*, L. from the first cleavage till the trochopore stage, with special reference to its 'chemical embryology', *Arch. neerl. Zool.* **7**, 353–434.

Raven, Chr. P. (1966), *Morphogenesis. The Analysis of Molluscan Development* (2nd Edition), Pergamon Press, Oxford.

Reverberi, G. (1971), 'Annelids' In G. Reverberi (Ed.) *Experimental Embryology of Marine and Fresh Water Invertebrates*, pp. 126–163, North Holland Publishing Company.

Seagrove, F. (1941), 'The development of the serpvlid *Pomatoceros triqueter*', *Quart, J. Mic. Sci.*, **82**, 467–540.

Smyth, J. D. (1962), *Introduction to Animal Parasitology*, London, English Universities Press.

Stephenson, J. (1930), *The Oligochaeta*, Oxford, Clarendon Press.

Taylor, E. R. and Baker, J. R. (1968), *The Cultivation of Parasites in vitro*, Oxford, Blackwell.

Welch, P. S. (1921), 'Bifurcation in the embryos of Tubifex', *Biol. Bull. Woods Hole*, **41**, 188–202.

4 Insects and Crustacea

Significant developmental studies in the arthropod phylum have been largely confined to insects and this chapter reflects this situation. We have deliberately ignored myriapods and arachnids and the crustaceans receive only brief attention. Many insects are easy to keep and they breed quickly in captivity. Under laboratory conditions their life histories may be studied easily and the cultures will provide a wealth of eggs and larval forms. An additional advantage is the existence, in several laboratory bred cultures of insects, of deleterious genes which affect all stages of development, including the formation of the egg itself. For teaching purposes, however, the nature of most insect eggs makes the early stages of development difficult to observe and understand. This is not true of the the larval forms, some of which have proved extremely useful for demonstrating certain aspects of post-embryonic growth and development. As an introduction to the study of insect development, and to give some idea of the kind of practical work which is feasible with this group, we have selected three insects for detailed treatment, namely, *Locusta* (the locust), *Calliphora* (the blow fly) and *Drosophila* (the fruit fly). Before describing them in detail we feel it would be helpful to place these examples in a more general context by giving a brief outline of the main features of insect development and then to indicate the significance of the practicals described. This introductory guide is necessarily brief. For comprehensive and more carefully qualified accounts of insect development reference should be made to standard texts and specialist reviews (e.g. Imms, 1957; Johannsen and Butt, 1941; Krause and Sander, 1962; Wigglesworth, 1965; Anderson, 1966; Counce and Waddington, 1972).

The eggs of insects are formed in paired ovaries, each of which is made up of a number of ovarioles. These are basically epithelial tubes in which eggs, generated at the anterior end, mature as they pass down the ovarioles on their way to the exterior. A common duct serves both sets of ovarioles and associated with this duct are accessory structures whose function is to ensure the fertilization of the eggs (where this is necessary) and to aid their survival once they are laid. In a few cases each ovary consists of only two ovarioles (e.g. in the ovo-viviparous Tse-tse fly) but in most cases there are many more (e.g. about 45 in *Locusta*, about 100 in *Calliphora* and about 200 in *Drosophila*). Two main types of ovariole are recognized; those in which the eggs develop in the absence of nurse cells (the panoistic type) and those in which nurse cells, also derived from the germinal epithelium, are present (the meroistic type). The location of these nurse cells in relation to the oocyte leads to a further sub-division; an acrotrophic type, in which the nurse cells remain associated with the growing oocyte as it passes down the ovariole, and a telotrophic type in which the nurse cells remain in the anterior part of the ovariole whilst maintaining a cytoplasmic connection with the receding oocyte. The number of nurse cells associated with an oocyte is constant for a given species. Thus in *Drosophila* (and other members of the family Muscoidea) four successive divisions of a terminal cystoblast (the progenitor cell) lead to the formation of 15 nurse cells and one oocyte. In meroistic ovarioles the oocyte grows at the expense of the nurse cells, which eventually degenerate. The protective chorion of the egg is fashioned by follicle cells, which are present in all types of insect ovary. For a review of ovarian structure, see Bonhag (1958).

With few exceptions the eggs of insects are sausage shaped and vary in length from about 0.5 to 5.0 mm. The egg is bounded by a cortex, beneath which is a clear cytoplasmic zone surrounding the centrally placed yolk. Typically, the eggs are protected by a chorion which often bears sculptured projections and is always pierced by a number of micropyles (openings for sperm entry) at the anterior (animal pole) end.

The posterior end is indicated by a clear cytoplasmic region, the posterior pole plasm, which often contains distinctive granules. With very few exceptions, development following fertilization proceeds by a series of nuclear divisions which result in a peripheral arrangement of nuclei in the cortical region. The formation of cell boundaries between the nuclei gives rise to a blastoderm, which quickly differentiates and produces recognizable insect features on its ventral surface. A typical example of insect development is illustrated in (Fig. 4.5). A few insects lay small eggs which undergo total cleavage. In some cases (Collembolids), this may reflect a primitive condition, but in others (some parasitic Hymenoptera), it appears to be a secondary adaptation related to yolk loss, and is associated with the evolution of polyembryony.

The experimental analysis of insect embryogenesis has been largely concerned with the relative roles of nucleus and cytoplasm, and with the regulative capacity of the developing embryo. The techniques used in such analysis have involved centrifugation of intact eggs and embryos, selective destruction of small areas of egg cytoplasm and ligaturing either the egg or early embryo in an attempt to prevent the free flow of material from one point to another.

Such techniques have also been used to investigate the role of the posterior pole plasm in the formation of germ cells (Duzynskaya, 1957) and to substantiate the ideas of activation and differentiation centres; also to distinguish between 'mosaic' and 'regulative' types of egg, where it has been shown, for instance, that the embryos of some species, (honey-bee) are able to regulate their development up to the blastoderm stage, whereas in others, (*Drosophila*) development appears to be strictly determined at fertilization. However, under certain conditions even the apparently strictly mosaic eggs of the Diptera seem to possess considerable powers of regulation (Yajima, 1960, 1964) and the application of the term 'mosaic' to these eggs needs careful qualification (Anderson, 1966).

The pattern of post-embryonic development varies from one insect to another but it is broadly divisible into two main types.

INSECTS AND CRUSTACEA 53

In hemimetabolous insects (e.g. *Locusta,* Fig. 4.1) the hatched embryo passes through a succession of instars, each accompanied by a moult, in which a gradual progression to the adult form is seen. In holometabolous insects (e.g. *Drosophila* Fig. 4.9 and *Calliphora* Fig. 4.6) there is a distinct larval phase which is quite unlike the adult form. In this case the growth of the larva is accompanied by a succession of moults culminating in a pupal phase during which a dramatic metamorphosis to the adult form occurs. Growth, moulting and metamorphosis are linked processes which are controlled by hormones, and the nature of this control is indicated in the practicals involving the larvae of *Drosophila* and *Calliphora*.

As mentioned above, the practical work described in this section makes use of three well known insects. In each case some essential details are given about the maintenance of

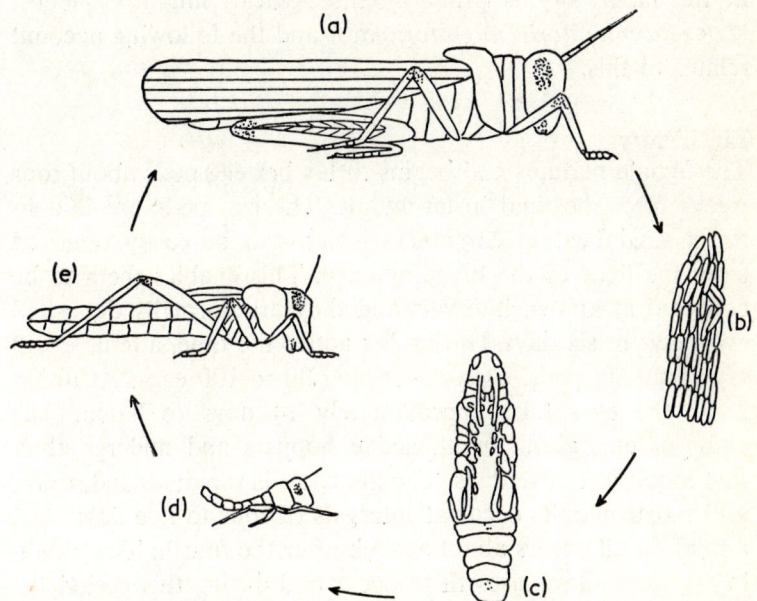

Fig. 4.1. *Stages in the life cycle of Locusta* (a) Adult. (b) Eggs laid in pod. (c) Embryonic development. (d) First hopper instar. (e) Fifth hopper instar.

cultures, together with a brief account of their life history. This is followed by more detailed information concerning practical work on particular aspects of development. *Locusta* is used to demonstrate the structure of the panoistic ovary and the external features of embryonic development; the practical using *Calliphora* concentrates on the role of hormones in post-embryonic development, and the work with *Drosophila* larvae draws attention to the relationship between chromosomal activity and developmental processes.

The Locust

The locust is now so commonly used for teaching purposes that there is no need to describe the method of breeding and maintaining a stock; relevant details will be found elsewhere (Hunter-Jones, 1961; Barras, 1964). The best species to keep in the laboratory is probably the African migratory locust (*Locusta migratoria migratorioides*) and the following account relates to this.

Life history

The female matures and begins to lay her egg pods about four weeks after the final instar moult. The egg pods are laid in moist sand held in containers which can be easily removed from the floor of the breeding cage. This enables them to be collected at known intervals and they are normally deposited every five or six days. During her active life time, a female will lay about six pods, each containing 30 to 100 eggs. At about 28°C the eggs take approximately 14 days to hatch. The embryos emerge as small, active hoppers and undergo their first moult after five days. The life cycle is rapid; second, third, and fourth moults occur at intervals of four to five days, and a final moult occurs about a week after the fourth. The moulting process allows growth to occur and during this period the nymphs, in contrast to the larvae of *Drosophila* and *Calliphora*, resemble the adult in many ways. The life cycle of the locust is illustrated in Fig. 4.1.

Examination of the ovaries
The ovaries are obtained from a mature female, freshly killed with a mixture (50:50) of chloroform and ether. To remove the ovaries, the abdomen is opened from the dorsal surface by making two longitudinal incisions on each side of the terga and two transverse incisions, one at the posterior end of the abdomen and the other across the thoracic region. The wings of the locust should be removed before the start of the dissection. The dorsal covering is then freed gently from the underlying viscera. The paired ovaries lie close together and are located on the dorsal side of the gut in the lower half of the abdomen. Starting from the anterior end, the ovaries are gently detached from neighbouring tissues and finally removed by cutting the oviducts at the point where they diverge round the hind gut in the direction of the cloaca. Fig. 4.2. illustrates the dissection, the disposition of the organs in the abdomen, and the general arrangement of the ovarioles in the ovary. For a detailed account of the internal anatomy of the locust reference should be made to Albrecht (1953).

Once they have been removed, the ovaries should be moistened with saline, spread out on a glass slide, and viewed under a dissecting microscope. At low magnification the ovarioles are obvious, and it is a simple matter to tease individual elements out by using dissecting needles and fine forceps. The preparation is completed by placing a separated ovariole on to another slide, adding a drop of saline, and gently lowering a thin cover slip over the tissue. The fresh preparation provides an excellent demonstration of the panoistic ovariole and will show a linear array of the stages of oogenesis, including a convincing view of chromosomes with a lampbrush structure (Kunz, 1967). Although a phase microscope will reveal most detail in these cells, an ordinary microscope will also give results if the light is adjusted carefully.

Development of the locust
If egg pods are collected at known intervals after oviposition, a representative series of locust embryos from an early blasto-

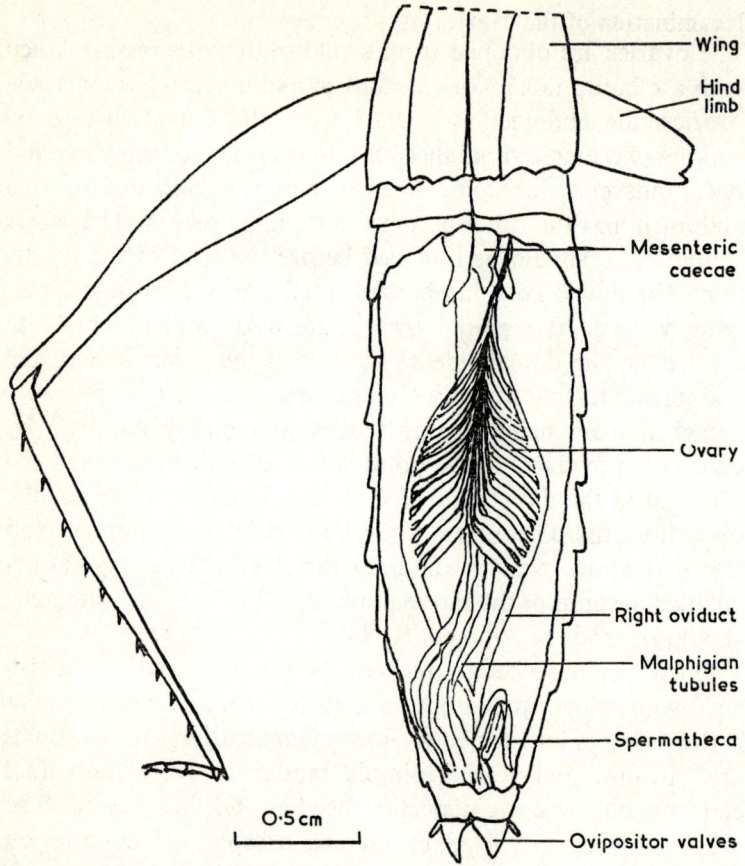

Fig. 4.2. Reproductive system of a female locust.

derm stage until hatching, can be obtained. There are usually 30 to 100 eggs per pod. The eggs are sausage shaped, about 6 to 7 mm long, and covered with an opaque white chorion. One end of the egg is blunt and permeated with pores (the hydropyle); it is also marked by the micropylar zone. The other end is more pointed. One side of the egg is slightly convex and the other almost straight. The external features of the egg corresponding to the ventral and dorsal surfaces of the embryo are illustrated in Fig. 4.3.

INSECTS AND CRUSTACEA

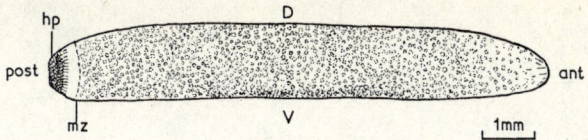

Fig. 4.3. *Locust egg*. The posterior end (post) is distinguished from the slightly more pointed anterior end (ant) by the presence of the micropylar zone (mz) and the hydropyle (hp). The ventral surface (V) of the egg is slightly concave and the dorsal surface (D) slightly convex.

In order to see the embryo the chorion of the egg must be removed. This can be done by dissolving the chorion in a dilute solution of sodium hypochlorite (Slifer, 1945). In practice we have found that a 0.5 per cent solution, made up in insect saline, will render the chorion transparent within five minutes, and reveal the yellow yolk of the egg and the white area of the embryo. After treatment the eggs should be washed carefully with insect saline. Alternatively the later stages of development, that is from three days onwards, can be obtained by disrupting the chorion mechanically. In this method, the protective coating of the egg is first perforated with fine needles at the end opposite to the micropylar zone and the chorion in this region completely removed. Some yolk will be expelled during this procedure, and the rest, together with the embryo, is pushed gently out of the remains of the egg case by means of a blunt seeker. The whole operation is carried out under insect saline, and is illustrated in Fig. 4.4. Permanent preparations of the embryos can be made by fixing them in 70 per cent ethanol, followed by staining in borax carmine.

The following is a brief account of the development of *Schistocerca gregaria*, staged according to Shulov and Pener (see Uvarov, 1966). The timing of the stages corresponds to a temperature of approximately 30°C. The development of this species appears to be very similar to that described for *Locusta migratoria* by Roonwall (1936, 1937).

The initial stages of development (stages I-IV), which take place during the first two days after the egg has been laid, are

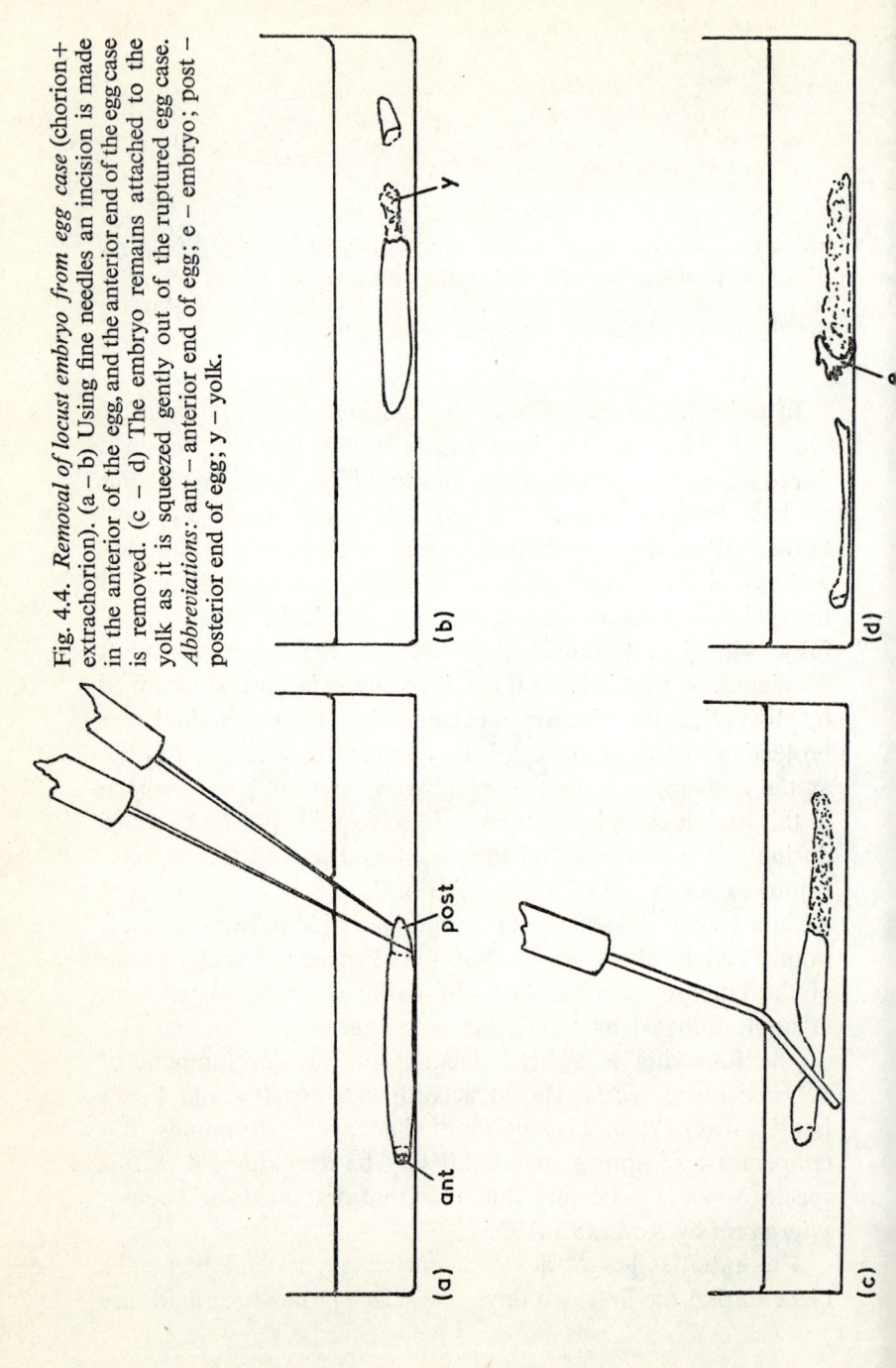

Fig. 4.4. *Removal of locust embryo from egg case* (chorion+extrachorion). (a – b) Using fine needles an incision is made in the anterior of the egg, and the anterior end of the egg case is removed. (c – d) The embryo remains attached to the yolk as it is squeezed gently out of the ruptured egg case. *Abbreviations*: ant – anterior end of egg; e – embryo; post – posterior end of egg; y – yolk.

not easy to distinguish. During this period the blastoderm is seen as a white, opaque patch located in the micropylar region of the yolk. By the end of the second day the blastoderm possesses a protocephalic region which can be distinguished from a narrower caudal region (stage V). Rudiments of the appendages develop during the course of the third day. At first they are equal in length (stage VII), but soon it becomes possible to distinguish the longer thoracic appendages and the shorter cephalic ones (stage VIII). During this period, when the embryo is between 2 and 3 mm long, it shifts slightly away from the micropylar region in a cephalic direction, to lie more ventrally. This so called 'anatrapetic movement' precedes the main 'blastokinetic movement' which shifts the embryo in the opposite direction, over the micropylar region, so that its ventral surface faces the dorsal side of the egg. This movement (characteristic of insects with large yolky eggs), occurs between the 7th and 8th day of development.

As development proceeds, the main features of the imago become more and more distinct. Thus, by the end of the fourth day, optic lobes, and mandibular, antennal and abdominal rudiments can be seen. At this stage relatively long thoracic rudiments become folded to touch in the mid-ventral line (stage XI). The increasing complexity is accompanied by growth in length, and by the seventh day, when individual mouth parts can be seen, the embryo is about 3.5 mm long (stage XIII). During the next three days the embryo grows rapidly, and by the tenth day it is nearly as long as the egg itself. By this time it is readily recognizable as an insect (stage XX), and spontaneous movements will occur if it is released from the egg case at this stage. The embryo hatches as a young hopper about 13 days after laying (stage XXIII). Fig. 4.5 illustrates some of the main features of the development of the locust.

Calliphora

The larva of the familiar blowfly, *Calliphora erythrocephala,* has proved extremely useful for demonstrating the hormonal

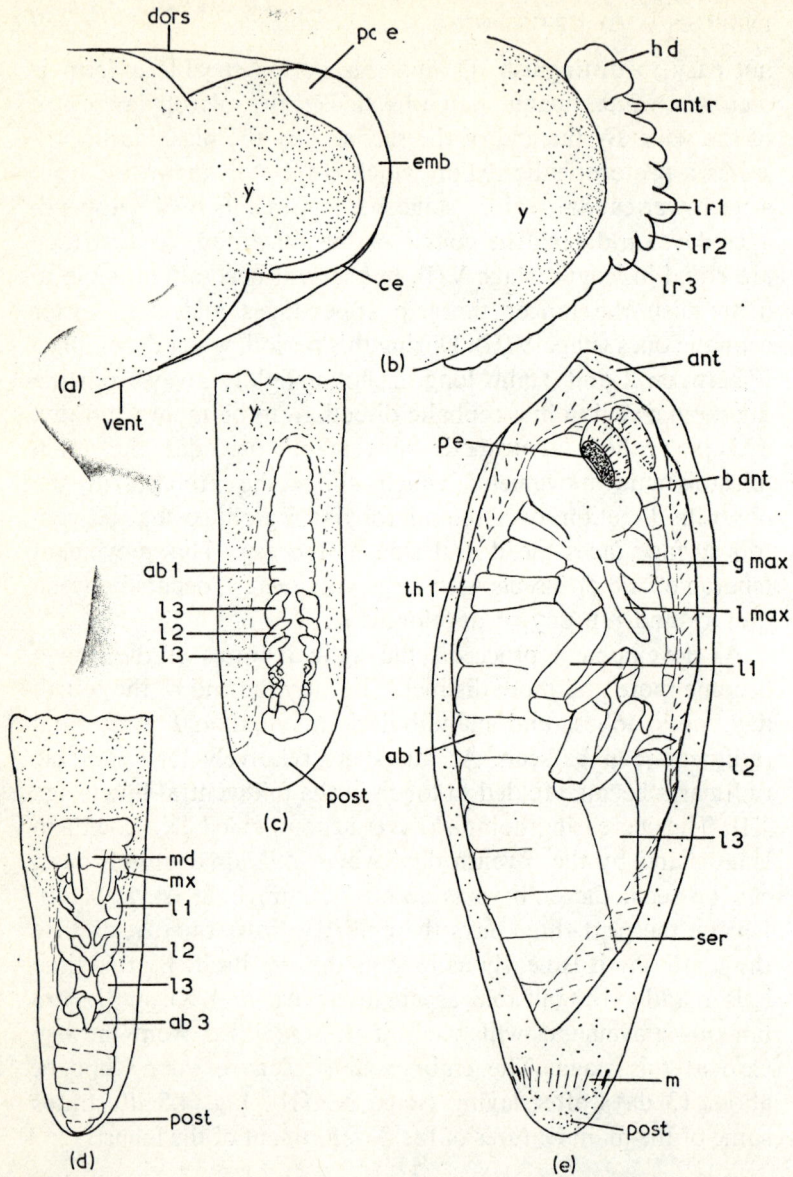

Fig. 4.5. *Stages in the development of the Locust* (a) 2 days (b) 3 days (c – d) 7 – 8 days Blastokinesis, (e) 10 days. *Abbreviations:* ab1 – first abdominal segment; ab3 – third abdominal segment; ant – anterior region of the egg; ant r – antennal rudiment; b ant – basal segment of antenna; ce – candal region of embryo; dors – dorsal surface of egg; emb – embryo;

control of pupation in Diptera, and a practical exercise based on this work is described in more detail below. As a pre-amble to this account, however, it seems worthwhile describing the breeding and maintenance of *Calliphora* in the laboratory, and giving an outline of its life history. The animal has a short generation time and its usefulness as teaching material is not confined to the larval stages.

Maintenance and life cycle
Of the several methods described for keeping and breeding *Calliphora* we find those conditions described by Scott (1934), to be adequate. The adult flies can be kept in a wooden framed enclosure walled and roofed with perforated zinc sheeting, and with an opening fitted with a sock shaped trap. They should be provided with water placed in small dishes (petri dishes) containing pieces of cotton gauze to prevent the flies drowning. The only source of food required for the adults are lumps of sugar. The cages should not be overcrowded; about a dozen flies per cubic foot is suggested. There is really no need to heat the enclosure provided the temperature does not drop below 15°C. Eggs are obtained by leaving a small piece of fresh liver in the adult enclosure for about 30 minutes. Once this piece has been removed it can be replaced by another, and so on. It is wise to discard the first piece of liver as it will probably contain a poor batch of eggs. Hatching is achieved by placing the liver containing the eggs in a petri dish kept humid with a piece of moist cotton gauze, and maintained at a temperature of about 23°C. The larvae will emerge within 24 hours and feed on the liver. The lid of the petri dish should be perforated with very small holes, sufficient to allow gaseous exchange, but not large enough to allow the larvae to escape from the

---

g max – galea of maxilla; hd – head of embryo; l max – lacinia of maxilla; lr 1 – 3 – rudiments of legs; l 1 – 3 legs; md – mandible; m – micropylar zone; mx – maxilla; pc e – protocephalic region of embryo; pe – pigment region of eye; post – posterior region of embryo; ser – serosa; th1 – first thoracic segment; vent – ventral surface of egg; y – yolk. (From original drawings of *Schistocerca gregaria* by Miss L. A. Moxon.)

chamber. The lid of the petri dish should be secured by two strong elastic bands, again to prevent the escape of the larvae. When the larvae hatch they are placed over a layer of silver sand about two inches thick with an additional supply of liver. A larger container, e.g. a plastic sandwich box, is necessary for keeping larvae at this stage, but the incubation temperature should remain the same. After feeding for about five days the larvae become sluggish and leave the food in order to pupate in the sand. The young flies begin to emerge after five or six days and this is the time to place the 'nursery' in the adult enclosure. Fresh liver is provided for a couple of days, but as soon as eggs are laid the liver is removed and replaced by the adult diet of lump sugar. The life cycle is summarized in Fig. 4.6. For further practical details see Ashby (1972).

The eggs of *Calliphora* are small (about 1.5 mm long) and sausage shaped (about 0.3 mm in diam.). Eggs are laid in clumps. These vary in size and may contain anything from a few to several hundred eggs. The eggs can be removed from

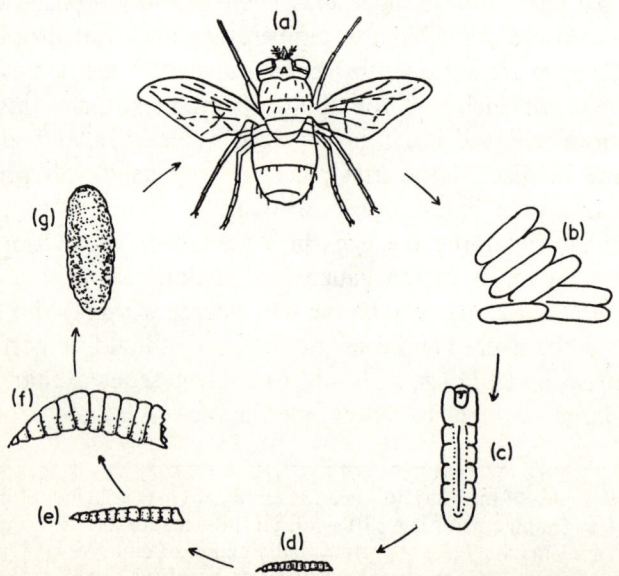

Fig. 4.6. *Stages in the life cycle of Calliphora* (a) Adult. (b) Eggs. (c) Embryonic development. (d) First larva. (e) Second larva. (f) Third larva. (g) Pupa.

the liver with fine forceps and a small paint brush, or by using a small loop of wire. Separation of individual eggs can be achieved manually, or alternatively the sticky mass can be separated by brief immersion in a 1 per cent solution of sodium hydroxide. Each egg is protected by a chorion, which can be removed by short exposure to a 0.5 per cent solution of sodium hypochlorite (see p. 57), followed by a thorough wash in insect saline. A full description of the embryology of *Calliphora* is given by Johannsen and Butt (1941).

Demonstration of hormonal control of metamorphosis in Calliphora (blowfly) larvae

Insect metamorphosis, like its counterpart in vertebrates, is under the control of hormones secreted by endocrine glands. It is most strikingly demonstrated in holometabolous insects (Diptera, Lepidoptera, Coleoptera) where, by means of a special moult, the larva is transformed first into a pupa and then into the adult winged form (imago). The sequence of events (Wigglesworth, 1965), is briefly as follows.

Neurosecretory cells in the protocerebrum secrete a hormone (prothorocotrophic or brain hormone) which acts on the ecdysial glands (also known as prothoracic glands in view of their location) causing them to secrete the moulting hormone, ecdysone. This acts synergistically with another hormone, secreted by the corpora allata, which is referred to as juvenile hormone or neotenin. When neotenin is in high concentration, the epidermal cells at the onset of moulting are caused to secrete a larval-type cuticle by the combined action of the two hormones. Prior to pupation when all the larval moults have been completed and the larva has reached its optimum size, the concentration of neotenin falls to a very low level and ecdysone now causes secretion of a pupal-type cuticle, as well as initiating other dramatic changes in the internal structure. In the absence of neotenin, ecdysone causes secretion of the adult cuticle, and since the ecdysial glands disappear during the transformation, no further moulting takes place.

In the higher Diptera, the endocrine glands (corpora allata,

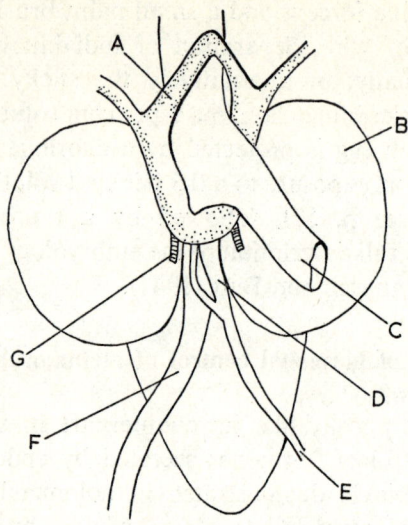

Fig. 4.7. *Weismann's ring in Calliphora.* A, Weismann's ring; B, brain; C, aorta; D, branch from recurrent nerve to Weismann's ring; E, recurrent nerve; F, gut; G, trachea to Weismann's ring. (From Wigglesworth, 1965).

corpora cardiaca and prothoracic glands) are fused together in the larva to form a ring of tissue surrounding the ventral aorta just anterior to the cerebral hemispheres (Fig. 4.7.). This composite endocrine organ is referred to as 'Weismann's ring' after its discoverer, or simply as the 'ring gland'. Fraenkel (1935) carried out simple ligature experiments on third instar *Calliphora* larvae and showed that only certain regions of the body pupated. This indicated that the ligatures had restricted the hormone to these regions. Fraenkel was not aware of the existence of Weismann's ring and implicated the ganglionic mass as the possible source of hormone. Practical details of the experiments are as follows.

Animals

Calliphora larvae which are one to three days post feeding are required. At this stage they have reached their maximal size and weight and are at the third instar. When next they moult they will transform into pupae. Larvae at this stage are very conveniently obtained from the Don Bait Company Ltd., Adwick Road, Mexborough, Yorks.

Ligaturing larvae

Weismann's ring can be found about one-third the length of the body from the anterior end. Active larvae which do not show signs of colouration are chosen, and treated as follows: a) Larvae (at least ten) are ligatured anterior to the approximate position of Weismann's ring. A loose knot is made in the ligature thread (ordinary white cotton will suffice) and the larva allowed to crawl through until it has reached the required position. The knot is then pulled tightly and secured with another knot so that the larva is constricted into two portions. b) Larvae are ligatured posterior to Weismann's ring. c) Larvae are ligatured posterior and anterior to Weismann's ring so that the body is divided into three segments. d) Larvae are left untreated to act as controls.

Experimental and control larvae are placed in labelled petri dishes, or some other suitable container, containing damp filter paper and left at room temperature.

Results

Daily examination should be made for signs of pupation and records kept of where and when this occurs. Examples of the results commonly obtained are depicted in Fig. 4.8. Some larvae may die due to the effects of the ligature, but the majority survive; others do not pupate, probably due to the ligature destroying the Weismann's ring. A large proportion will show complete pupation despite the ligature (d-e), and this is due to the fact that ecdysone had already reached the target organs before the ligature was applied. Successful experiments (a-c) are quite dramatic, showing coexisting larval and pupal (f) characteristics. However tanning of the cuticle may not take place despite the fact that the epirdemis is exposed to ecdyson. The reason for this is that oxygen is required for the chemical

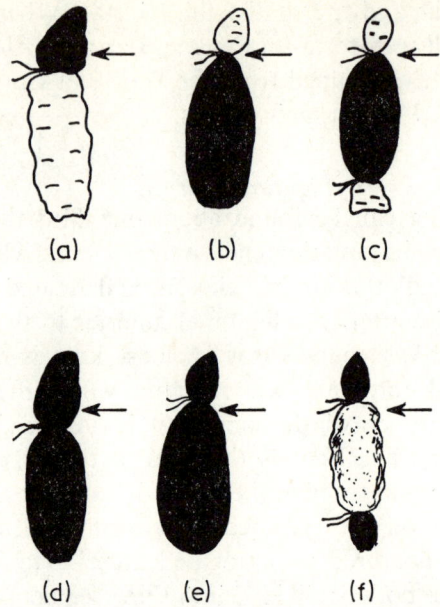

Fig. 4.8 Diagrams showing the appearance of *Calliphora* larvae after applying ligatures in various positions relative to Weismann's ring. Shaded areas denote formation of pupal type cuticle. Arrows denote position of Weismann's ring. (For explanation of differences, see text).

reactions to take place and the ligatures may completely constrict the tracheae leading from the anterior and posterior spiracles. The dependence on oxygen can be further demonstrated by blocking the anterior or posterior spiracles with vaseline as well as ligaturing the larvae. Those segments which have not pupated, due to the ligature restricting passage of the hormone, can be induced to pupate by injecting fluid from pupating larvae.

It is now known that ecdysone causes pupation by influencing the way in which the epidermal cells metabolize tyrosine. (Karlson, 1965). In early third instar *Calliphora* larvae, tyrosine undergoes a process of deamination leading to *p*-hydroxyphenylpyruvic acid, but in late third instar larvae tyrosine is metabolized along a pathway leading to

N-acetyldopamine quinone. This acts as the tanning agent, causing cross linking of the protein chains in the cuticle and consequently its hardening and colouration. One of the enzymes involved in the pathway is dopa-decarboxylase, and it is believed that ecdysone in some way causes a derepression of the genes in the epidermal cells so that mRNA specific for dopa-decarboxylase is encoded by the derepressed DNA.

Drosophila

Fruit flies (*Drosophila melanogaster* and related species) are commonly used in teaching laboratories for experiments in genetics and therefore a plentiful supply of embryos and larvae should be available for developmental studies. In view of this, and of the existence of a great deal of information on the biology and general maintenance of these animals (e.g. Demerec, 1950; Wheeler, 1972), the following notes have been kept to a minimum, with the specific requirements of the practical on polytene chromosomes in mind.

Life cycle and culture of larvae
The life cycle of *Drosophila* is short. At approximately 21°C fertilized eggs hatch in about 24 hours to form the first larval instar. These moult to form a second instar after a further 24 hours. A second moult to form the third instar occurs after a similar period of time. During the course of the next two days the third instar grows in size and finally pupates about 96 hours after hatching. The imago emerges about four days after pupation and oviposition usually occurs during the second day of adult life.

Stock cultures of larvae can be maintained on various nutrient media suitable for the growth of yeast on which the larvae feed (for various recipes see Wheeler, 1972). The following procedure is the one we use.

72 g of oatmeal (fine or medium grade) are allowed to soak overnight in 12 ml of water. 6 g of agar powder is dissolved in 400 ml of boiling water contained in a saucepan to which

is then carefully added 6 ml of a 10 per cent solution of Nipagin (a fungicide) in absolute alcohol. The oatmeal solution is then added and the solution brought to the boil with constant stirring. 35 g of treacle (Fowlers Black) is diluted with 38 ml of water and added to the boiling mixture. Boiling is allowed to continue for ten minutes. Whilst hot, the mixture is then poured to a depth of 2 cm in sterilized, half pint milk bottles. The quantities stated provide sufficient medium for 10 bottles. Each bottle is stoppered with non-absorbent cotton wool, allowed to cool, and stored at 4°C until required.

The day before adults are introduced, three drops of baker's yeast suspension are added to each bottle and these are then incubated at 22 to 24°C. A folded filter paper is pushed into the medium and serves to absorb excess moisture.

Eggs are laid on the surface of the medium and 1st, 2nd and 3rd larval instars feed voraciously on it. Late 3rd instar larvae will be found moving actively on the sides of the bottle, but at the onset of the prepupal moult these larvae become less active and at the point of moulting the anterior spiracles are rapidly everted. Prepupae have everted anterior spiracles, a more rotund shape, and a more translucent cuticle (see Fig. 4.9).

Observation of puffing patterns in the polytene chromosomes of Drosophila larvae

Cells in the salivary glands and certain other organs of larval Dipteran insects, such as *Drosophila* and *Chironomus*, have long been known to be very large compared to those in other parts of the larvae, and to contain exceptionally large chromosomes. These are referred to as 'polytene chromosomes' and they arise as a result of repeated DNA replication. A unique situation occurs in these organs in that the chromosomes are visible in interphase cells. When appropriately stained the chromosomes are seen to have very characteristic banding patterns (Plates 4.1 and 4.3). Careful analysis of these bands in relation to extensive breeding experiments that have been carried out on *Drosophila*, has indicated that the bands bear a direct relation-

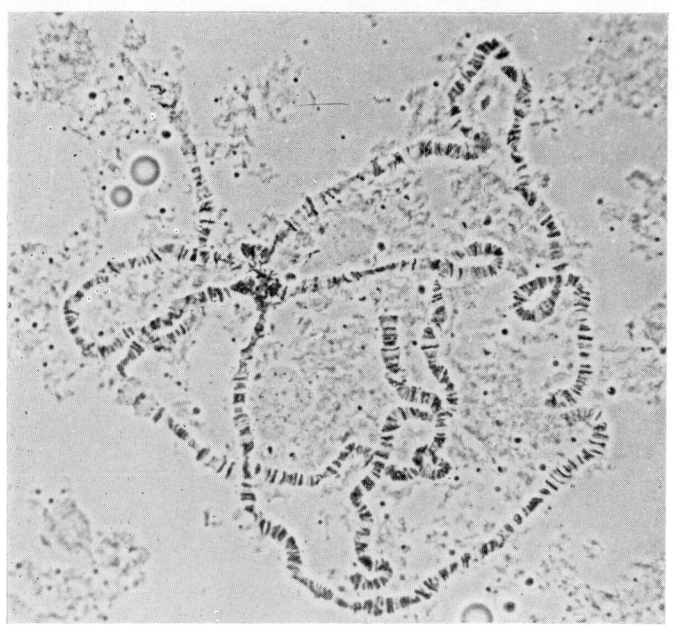

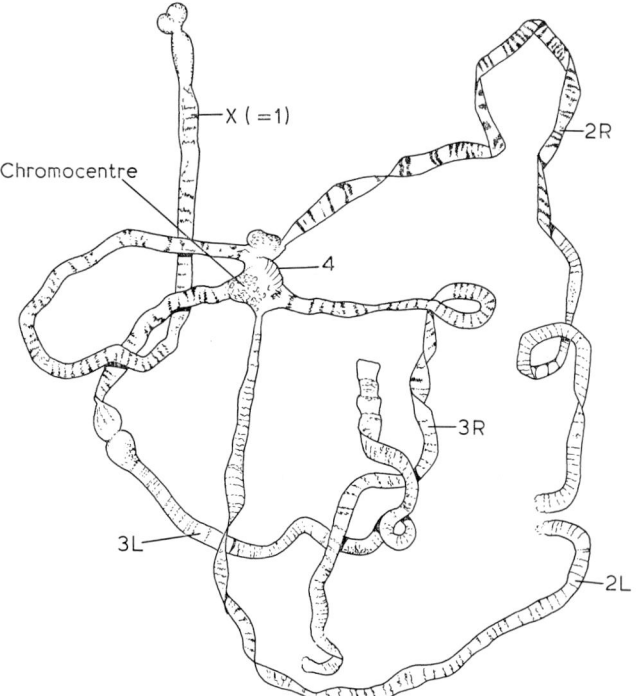

Plate 4.1 and accompanying explanatory figure.
Drosophila melanogaster salivary gland cell preparation from a late 3rd instar larva, showing the appearance of the four polytene chromosomes after squashing and staining. R and L denote right and left limbs of the large autosomes 2 and 3. (× 400)

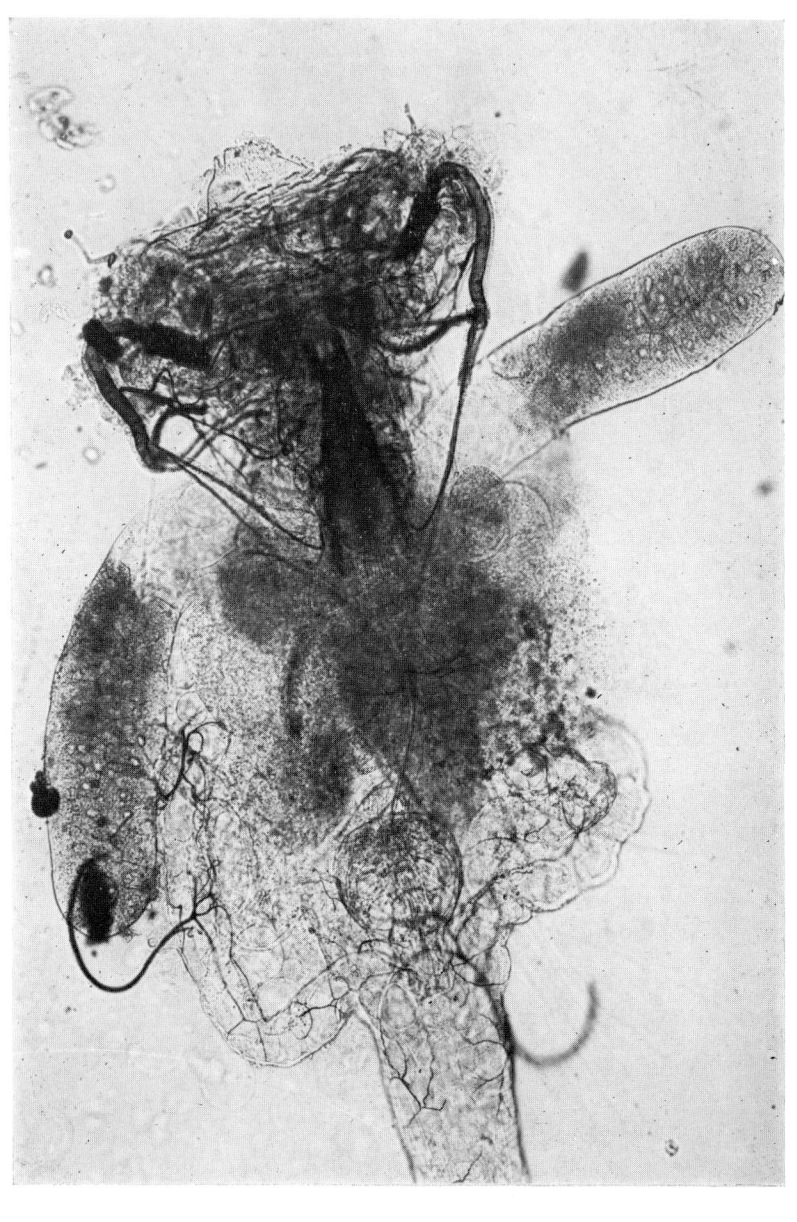

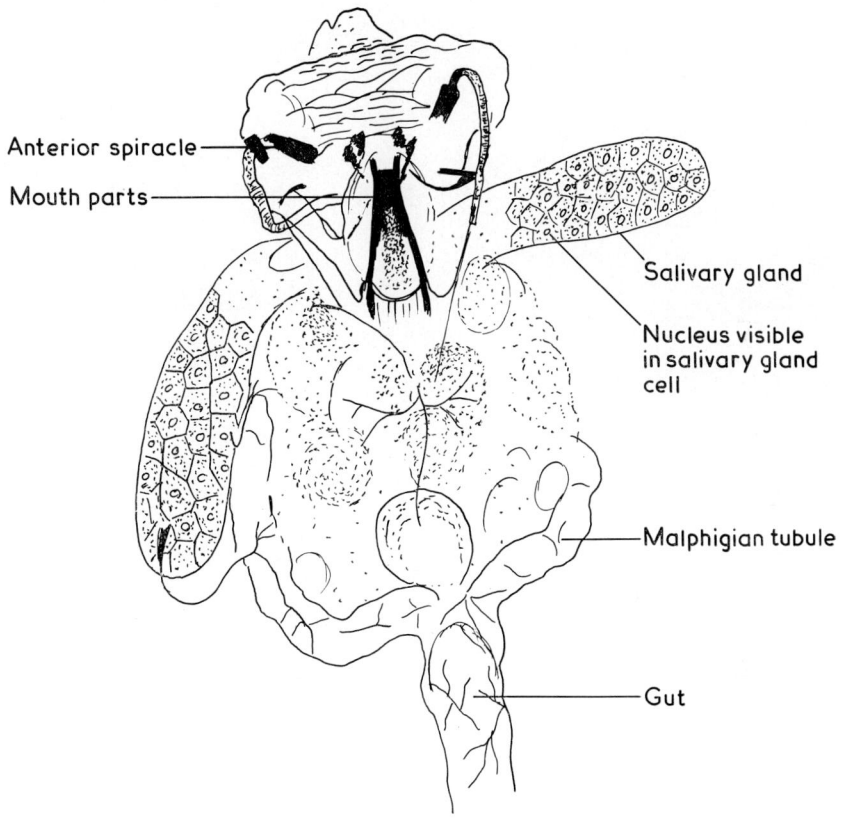

Plate 4.2 and accompanying explanatory figure
Appearance of the salivary glands and associated anterior structures in a late 3rd instar *Drosophila melanogaster* larva that has been pulled apart and viewed under a dissecting microscope. (× 40)

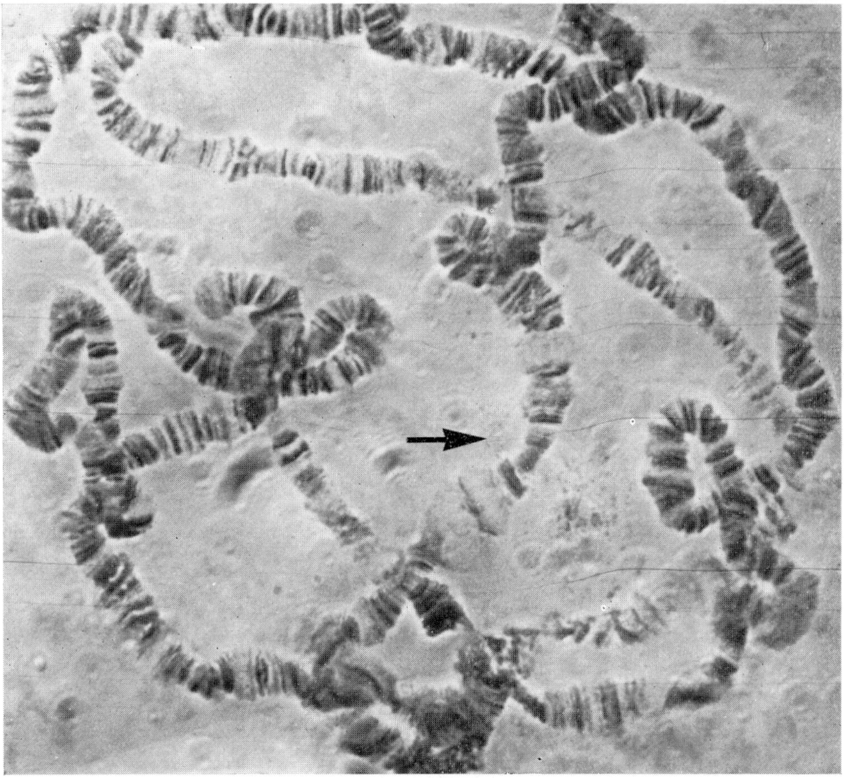

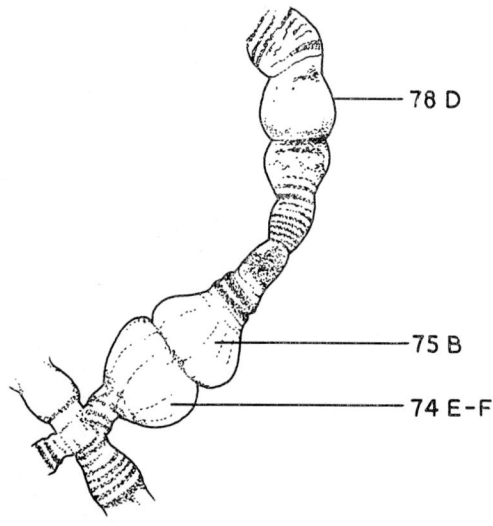

Plate 4.3 and accompanying explanatory figure
High power view of the polytene chromosomes in a salivary gland cell preparation from a late 3rd instar *Drosophila melanogaster* larva. The chromosomes show the characteristic banding pattern and stage specific puffs on chromosome 3L at loci 74EF, 75B and 78D (arrowed and redrawn in the accompanying figure). (× 1000)

INSECTS AND CRUSTACEA 69

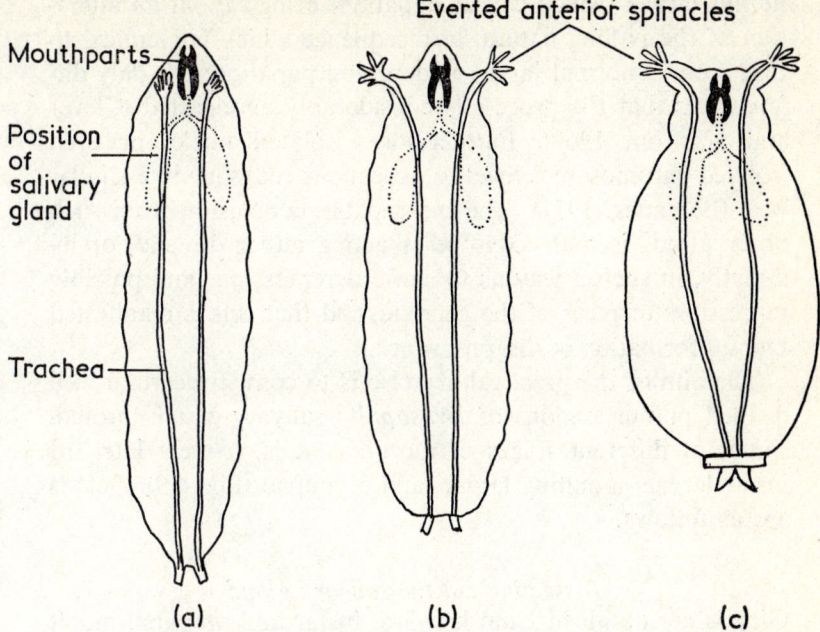

Fig. 4.9. Diagrams depicting differences in appearance of (a) late 3rd instar larva; (b) moulting larva; (c) prepupa, of *Drosophila melanogaster*.

ship to the genes (Sinnott, Dunn and Dobzhansky, 1952). A further feature of polytene chromosomes, and the one with which this practical exercise is primarily concerned, is the presence of characteristic 'puffs'. These are regions on the chromosomes where the DNA has become loosened into numerous outward looping strands and it is now known that messenger RNA is rapidly synthesized on these strands. Of considerable interest has been the finding that the puffs change their size and linear distribution during different stages of development. The puffing pattern has been carefully worked out for the salivary gland chromosomes of *Chironomus* (Beerman and Clever, 1964), and *Drosophila* (Becker, 1962; Ashburner, 1967) in relation to the morphogenetic changes consequent upon metamorphosis. Injection of ecdysone into *Chiro-*

nomus larvae (which causes pupation) brings about an alteration of the puffing pattern in a sequence which is identical to that seen in normal larvae undergoing pupation; the only difference is that the process is considerably accelerated (Clever and Karlson, 1960). Furthermore, isolated nuclei or even isolated chromosomes react to exogenous ecdysone in a similar way (Berendes, 1971). The inescapable conclusion from such observations is that ecdysone is acting either directly, or indirectly, in such a way as to cause derepression (and possibly repression) of parts of the genome, and that this is manifested by puff formation or disappearance.

The aim of the practical exercise is to compare certain well defined puffing regions of *Drosophila* salivary gland chromosomes at different stages of morphogenesis, namely late 3rd instar larvae, moulting larvae, and prepupae (Fig. 4.9). Details are as follows.

Dissecting out the salivary gland

Glands are obtained from late 3rd. instar and prepupal moult larvae in exactly the same way. Individual larvae are selected from the sides of the culture bottle with the aid of a fine brush and transferred to a drop of 0.4 per cent NaCl on a microscope slide. The slide is transferred to the stage of a dissecting microscope and the larva brought into focus. A dissecting needle is applied behind the mouthparts and another to the posterior region; the needles are then jerked apart so that the anterior section, together with the salivary glands, is detached from the rest of the body. Plate 4.2 shows the appearance of the salivary glands amongst the associated viscera at this stage. The salivary glands are paired structures united by their secretory ducts; fat bodies are often associated with them. The nuclei are easily seen in the large, polygonal, salivary gland cells (Plate 4.2). The glands are dissected free from the adherent tissue and transferred by means of a dissecting needle to a clean, dry microscopic slide.

For prepupae, a cut is made with fine scissors in the body wall in front of the mouth parts. Pressure is then applied at the anterior end and the viscera forced out. The salivary glands are then located and freed by dissection.

Staining and squashing the glands

Lactic propionic acid orcein stain is prepared as follows. 5 g of synthetic orcein (Gurr's) are dissolved in a mixture consisting of 50 ml of propionic acid (45 per cent in water) and 50 ml of lactic acid (45 per cent in water) by boiling in a reflux condenser for two days. The solution is filtered whilst still warm and stored at 4°C. Should precipitation occur the stain is further diluted with the acid mixture.

A drop of lacto-propionic acid orcein stain is applied to the gland so that it is covered (care must be taken to ensure that the gland does not at any stage dry out). Staining is allowed to continue for about eight minutes, whereupon a coverslip is carefully lowered onto the specimen taking care to avoid trapping air bubbles. The nuclear membrane of the salivary gland cells is broken by lightly tapping the coverslip immediately over the gland, and the chromosomes are then spread by squashing. This is achieved by placing a filter paper over the coverslip and applying as much vertical pressure as is possible with the thumb. Nail varnish is applied to the edge of the coverslip to prevent the preparation from drying out; such preparations can be stored for several days at 4°C.

Observations

A microscope with an oil immersion objective and preferably a phase attachment is essential for the analytical work required. Very good preparations will show the four chromosomes (actually homologous pairs tightly associated) attached by their centromere regions in a mass which shows less discernible banding and which is called the chromocentre (Plate 4.1). One chromosome (referred to as 4) is very short and has the centromeres at the ends; consequently it is often very hard to distinguish. The X chromosome (also referred to as 1) is much longer but also has the centromeres at the ends of the homologous pairs. Chromosomes 2 and 3 are very long and have their centromeres about midway, so that right (R) and left (L) limbs are produced. Regions of the chromosomes sometimes become dissociated into the homologous pairs during preparation of squashes. The chromosomal regions are identified by reference to the maps constructed by Bridges, copies of which are shown in Sinnott, Dunn and Dobzhansky (1952). These maps show the thickness, relative staining intensity, and linear distribution of the bands in coded form.

Puffs will be recognized as localized increases in chromosome diameter (the larger ones are sometimes referred to as 'Balbiani rings' after their discoverer) and as many as 108 loci on chromosomes 2 and 3 have been reported to show puffs at some time during the late larval and prepupal stages (Ashburner, 1967). It will be apparent that the puffs and banding patterns are identical for the same chromosomes originating from different cells in the salivary gland. Particular attention should be paid to regions 78 D, 75 B, 74 EF, and 71 CE on chromosome 3L, and 85 D, 85 EF on chromosome 3R. These regions, especially those on chromosome 3L, are not too difficult to identify. Puffs should be scored as absent $= 0$, small $= 1$, medium $= 2$ and large $= 3$, and plotted in histogram form for each of the three stages (Fig. 4.10). It will be observed that the puff at locus 78 D is quite large in the late third instar stage, much smaller in the moulting larva, and absent in the prepupa. Puffs at loci 75 B, 74 EF, which are very large in the late third instar stage, do not completely disappear in the prepupa; on the other hand the puffs at loci 71 CE and 85 D appear for the first time in the prepupa.

It is not easy to show a direct link between ecdysone, puffing in salivary gland chromosomes, and metamorphosis, since the morphogenetic changes are largely manifested in epidermal cells which secrete pupal and imaginal type cuticles, rather than in salivary gland cells. However, evidence for a relationship can be obtained by following a procedure described by Becker (1962), and for more advanced students this makes an interesting extension of the practical.

Late third instar larvae are ligatured so that the salivary gland is divided into an anterior and posterior half. This is

INSECTS AND CRUSTACEA

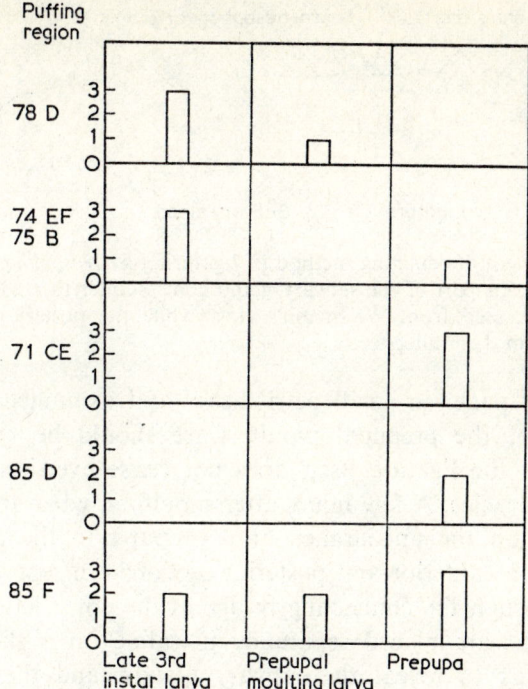

Fig. 4.10. Histograms showing differences in size and presence of puffs at loci 78D, 74EF, 75B, 71CE, 85D, 85F in late 3rd instar larva, prepupal moulting larva, and prepupa of *Drosophila melanogaster*. 0 = absent, 1 = small, 2 = medium and 3 = large.

accomplished by making a loose knot in a hair (from a baby) and securing one end in plasticine on a microscope slide. The larva is pushed into the loop with forceps and positioned so that a tight knot can be tied at about the fourth segment. Weismann's ring (see p. 64) which lies anterior to the supra-oesophageal ganglia, should now lie in the anterior region together with a portion of the salivary gland (Fig. 4.11). If the ligature has been applied at the right time, that part of the animal anterior to the ligature will come under the influence of ecdysone and pupate; the posterior part of the animal, which also contains a portion of salivary gland, will remain larval. Larvae ligatured in this way are placed separately on

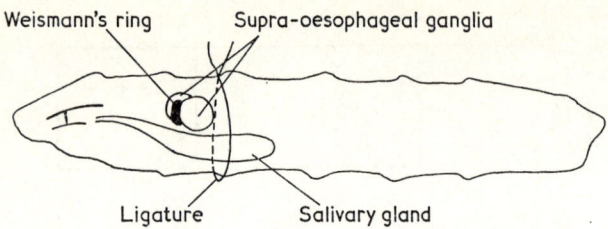

Fig. 4.11. Diagram depicting method of ligaturing a *Drosophila* larva so that the anterior part of the salivary gland comes under the influence of hormones secreted from Weismann's ring, while the posterior part is protected from their influence.

damp filter paper in small petri dishes and examined hourly for signs of the prepupal moult. Care should be taken to ensure that the ligature itself does not cause eversion of the anterior spiracles. A few hours after moulting, when the larva has taken on the appearance of a prepupa in the anterior half only, the anterior and posterior sections are separated by cutting through the connecting bridge at the constriction. The two sections are placed separately in saline on a slide, dissected so as to reveal the salivary glands, and the glands stained and squashed as previously described. A record is then made of the presence or absence of puffs at the same loci on chromosomes from cells of the anterior and posterior portions.

Recent work (Whitten, 1968) on the foot pad cells of the flesh-fly *Sarcophaga bullata*, has shown that they too contain giant polytene chromosomes. These cells are intimately concerned in the secretion of the imaginal type exoskeleton characteristic of the foot. They undergo changes in puffing patterns during differentiation of the foot, but it is not yet known whether the sequence of changes is the same as that in the salivary gland.

Crustacea

It is not easy to study the early development of crustacea, because many species have a strictly seasonal production of eggs. Even when they are available the eggs are not especially useful as they are enclosed in a tough opaque shell. Thus, although it is quite common at certain times of the year to see female crabs and crayfish carrying egg masses, little can be done with such material except to observe the hatching of the young. Possibly the best, or at least the most readily available, sources of embryological material are those species in which the young remain protected in the female, for instance in the brood pouch of *Daphnia* or the 'marsupium' of wood lice (*Oniscus* species). Because of the difficulties associated with studying the embryology of this group, it is inevitable that more attention is paid to the post-embryonic stages, that is to the succession of larval forms which lead, by a series of moults, to the adult. Details of some of the earlier work on the development of crustacea will be found in MacBride (1914), and more recent accounts are given by Green (1965, 1971).

Larval forms
Some crustacea pass through larval stages during which new segments and appendages are added successively until the young adult stage is reached. This type of transition, in which no dramatic change of form occurs as the larva changes to the adult, is sometimes spoken of as *anamorphic,* in contrast with more dramatic changes which may occur during the life history, and which are known as *metamorphic.* Alternatively, a distinction may be drawn between *monophasic* life histories, where there is no sudden change in form between one moult and the next, and *di, tri,* and *polyphasic* life histories, in which there is one or more striking changes in form (Clark, 1973). These changes are obviously related to the different habitats which the larval and adult animals occupy. For instance, the typical zooea larva of the crab is adapted to life amongst the plankton, whereas the adult is adapted to living on the sea bed. More extreme changes of form occur in parasitic forms such as *Sacculina.* Some of the crustacean larval forms resemble the free-living adults of other species and are named accordingly. Hence, the *nauplius* stage is identical in appearance

with *Nauplius* itself, and the *cypris* stage of barnacles resembles *Cypris*. Many of the typical crustacean larval forms can be demonstrated in plankton samples collected at the right time of the year and others can be obtained easily from laboratory cultures of small crustacea, such as *Daphnia* and *Artemia*. Details of the culture of many Crustacea are given in Galtsoff *et al.* (1959). Our own experience has been with *Artemia*.

The culture of Artemia salina
It is relatively easy to maintain cultures of brine shrimps in the laboratory. These not only provide a readily available source of food for fish and urodele larvae, but also a convenient source of metanauplii for teaching purposes. The time of hatching, rate of growth and the length of the life cycle, obviously depend on the temperature; the times given below correspond to an average laboratory temperature of 20°C.

To obtain a culture of brine shrimps, eggs (obtained from a local aquarist) should be sprinkled gently onto the surface of about 700 ml of artificial sea water contained in a 1000 ml beaker. Sufficient eggs are needed to form an even layer on the surface of the water. Some of the eggs will sink to the bottom of the beaker, but the majority will remain at the surface layer. The water need not be aerated, and we usually leave the beaker undisturbed for at least a week. After 2 to 3 days, the *Artemia* emerge as a metanauplius stage and they appear in increasing numbers during the following week. After about 10 days the young larvae must be decanted away from the unhatched eggs into another beaker containing clean artificial sea water, and if they are to be kept alive they will need feeding. A convenient source of food is Complan, and a periodic addition of a few drops of a 5 per cent w/v suspension enables the animals to grow and reach sexual maturity in five or six weeks. It is important not to overfeed and fresh Complan suspension must only be added when the water is clear. If the water becomes fouled the *Artemia* should be transferred to clean artificial sea water. Sexual maturity is indicated by mating and the subsequent appearance of egg sacs on the

INSECTS AND CRUSTACEA 77

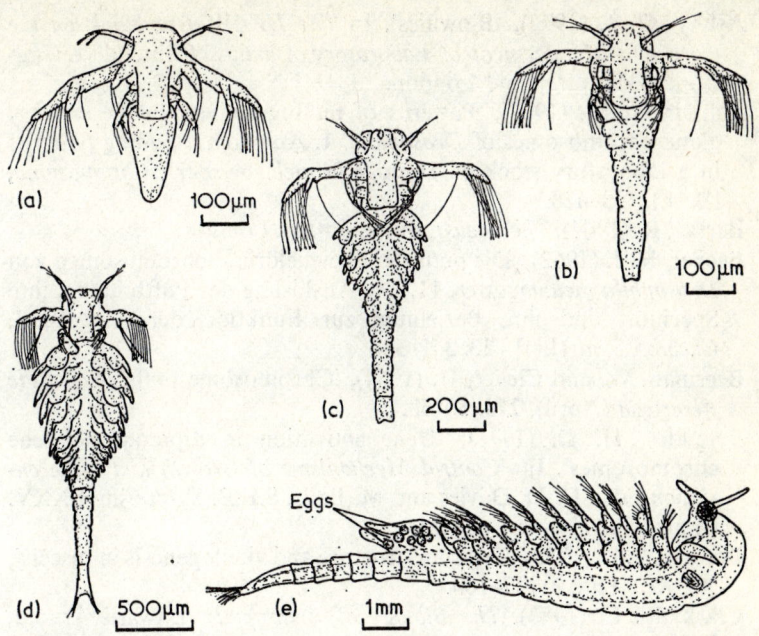

4.12. *Stages in the life cycle of Artemia* (a) Metanauplius, first instar. (b) Second instar. (c) Fourth instar. (d) Seventh instar. (e) Adult female. (a – d, after Heath, 1924; e, after Green, 1961).

females. Within a week of mating many eggs are shed onto the bottom of the container and soon after this the next generation of metanauplii appear. A new culture is set up by removing the adults with a wide mouthed pipette and decanting the young larvae into clean artificial sea water, away from unhatched eggs and accumulated debris. The life cycle of *Artemia* is summarized in Fig. 4.12. For class purposes the larvae should first be examined alive and then in more detail after fixing in glacial acetic acid fumes.

REFERENCES

Anderson, D. T. (1966), 'Comparative embryology of Diptera', *Ann. Rev. Ent.* **11**, 23–46.

Albrecht, F. O. (1953), *The Anatomy of the African Migratory Locust*, University of London, Athlone Press.

Ashby, G. J. (1972), 'Blowflies'. In *The UFAW Handbook on the Care and Management of Laboratory Animals*. Churchill Livingstone, Edinburgh and London.

Ashburner, M. (1967), 'Patterns of puffing activity in the salivary gland chromosomes of *Drosophila*. 1. Autosomal puffing patterns in a laboratory stock of *Drosophila melanogaster*', *Chromosoma*, (Berl.), 398–428.

Barras, R. (1964), *The Locust*. Butterworths, London.

Becker, H. J. (1962), 'Die puffs der Speicheldrüsenchromosomen von *Drosophila melanogaster*. 11. Die Auslösung der Puffbildung, ihre Spezifitat und ihre Beziehung zur Funktion der Ringdruse', *Chromosoma*, (Berl.) **13**, 341–384.

Beerman, W. and Clever, U. (1964), 'Chromosome puffs', *Scientific American* (April), **210**, 50–58.

Berendes, H. D. (1971), 'Gene activation in dipteran polytene chromosomes'. In: *Control Mechanisms of Growth and Differentiation* Eds., D. D. Davies and M. Balls. S.E.B. Symposium XXV. Cambridge University Press.

Bonhag, P. F. (1958), 'Ovarian structure and vitellogenesis in Insects', *An. Rev. Ent.*, **3**, 137–160.

Clark, K. E. (1973), *The Biology of Arthropods*. Edward Arnold, London.

Counce, S. J. (1961), 'The analysis of insect embryogenesis', *Ann. Rev. Ent.* **6**, 295–312.

Counce, S. J. and Waddington, C. H. (Eds) (1972), *Developmental Systems: Insects*, Volumes 1 and 2, Academic Press, London and New York.

Clever, U. and Karlson, P. (1960), 'Induktion von Puff-Veränderungen in den Speicheldrüsenchromosomen von *Chironomus tentans* durch Ecdyson'. *Exp. Cell Res.* **20**, 623–626.

Demerec, M. (Ed) (1950), *Biology of Drosophila*, Wiley, New York and London.

Duszynskaya, I. G. (1957), 'Experimental research on chromosome elimination in ceccidomyidae (Diptera)', *J. exp. Zool.*, **141**. 391–441.

Fraenkel, G. (1935), 'A hormone causing pupation in the blowfly *Calliphora erythrocephala*'. *Proc. Roy. Soc. B*, **118**, 1–12.

Galtsoff, P. S., Lutz, F. E., Welch, P. S. and Needham, J. G. (1937), *Culture Methods for Invertebrate Animals*. Republished in 1959 by Dover Publications Inc., New York.

Green, J. (1961), *A Biology of Crustacea*, Witherby, London.

Green, J. (1965), '*The chemical embryology of Crustacea*', Biol. Rev. Cambridge Phil. Soc., **40**, 580–600.

Green, J. (1971), 'Crustaceans', In Reverberi, G. (Ed). *Experimental Embryology of Marine and Fresh water Inverterbates*, North Holland Publishing Company, Amsterdam and London.

Heath, H. (1924), 'The external development of certain Phyllopods' *J. Morph.*, **38**, 453–475.

Hunter-Jones, P. (1961), *Instructions for Breeding and Rearing Locusts in the Laboratory*, Anti-locust Research Bulletin: London.

Imms, A. D. C. (1957), *A General Textbook of Entomology*, Ninth Edition. Revised by O. W. Richards and R. G. Davies. London.

Johannsen, O. A. and Butt, F. H. (1941), *Embryology of Insects and Myriapods*, McGraw-Hill, New York.

Karlson, P. (1965), 'Biochemical studies of ecdysone control of chromosome activity'. *J. Cell Comp. Physiol.* **66**, 69–76.

Krause, G. and Sander K. (1962), 'Ooplasmic reaction systems in Insect embryogenesis'. *Advances in Morphogenesis* **2**, 259–303.

Kunz, W. (1967), 'Funktionstrukturen in Oocytenken von *Locustus migratoria*'. *Chromosoma* **20**, 332–370.

MacBride, E. W. (1914), *Text-Book of Embryology*, Volume I *Invertebrata*, Macmillan, London.

Roonwall, M. L. (1936), 'Studies of the embryology of the African migratory locust, *Locusta migratoria migratorioides* R & F., I The early development with a new theory of multiphased gastrulation among insects', *Phil. Trans. Roy. Soc.*, (B), **226**, 391–421.

Roonwall, M. L. (1937), 'Studies on the embryology of the African migratory locust, **II** Organogeny'. *Phil. Trans. Roy. Soc.* (B) **227**, 175–241.

Scott, C. M. (1934), 'Action of X-rays on *Calliphora*. With appendix on method of preparing eggs for microscopy by Kilgour'. *Proc. Roy. Soc. B*, **115**, 100–121.

Sinnott, E. W., Dunn, L. C. and Dobzhansky, T. H. (1952), *Principles of Genetics*. Fourth Ed. McGraw-Hill, New York.

Slifer, E. H. (1945), 'Removing the shell from living grasshopper eggs', *Science*, **102**, p. 282.

Uvarov, B. (1966), *Grasshoppers and Locusts*, Volume I, Cambridge University Press, Cambridge.

Wheeler, M. R. (1972), 'Fruitflies'. In: *The UFAW Handbook on the Care and Management of Laboratory Animals.* Churchill Livingstone, Edinburgh.

Whitten, J. (1968), 'Metamorphic changes in insects'. In: *Metamorphosis – a problem in developmental biology.* pp. 43–105. Etkin, W. and Gilbert, L. I. (Eds). North Holland Publishing Company, Amsterdam.

Wigglesworth, V. B. (1965), *The Principles of Insect Physiology*, Chapman and Hall, London.

Yajima, H. (1960), 'Studies on embryonic determination of the harlequin-fly, *Chironomus dorsalis*, I, Effects of centrifugation and of its combination with constriction and puncturing'. *J. Embryol. exp. Morph.*, **8**, 198–215.

Yajima, H. (1964), 'Studies on embryonic determination of the harlequin fly, *Chironomus dorsalis*. II, Effects of partial irradiation of the eggs by ultra-violet light', *J. Embryol. exp. Morph.*, **12**, 89–100.

5 Fish

Apart from a quick look at fairly well-advanced trout embryos the study of fish development is often neglected in embryological courses in the United Kingdom. This is not really surprising in view of the emphasis which is placed in many text books on the descriptive embryology of other vertebrate types, invariably amphibia and chick, and of the relatively minor contribution that work on fish embryos has made to the causal analysis of development. However, in our experience, fish, and especially teleosts, possess certain advantages over the more familiar material for certain types of practical work on vertebrate embryos. Above all, the embryos of certain tropical fish demonstrate very clearly the main features of the blastodermic type of development and the development of the major organ systems of vertebrates. These fish have the additional advantage that they can be maintained and bred with ease in the laboratory, so producing a continuous supply of eggs throughout the year. We believe that this material should be placed high on the list for an introductory course in embryology. Despite the enthusiasm of a few contemporary embryologists (e.g. Cohen, 1967), it is regretted that teleosts are not yet widely used for teaching some of the basic facts of development.

In this short chapter we have concentrated on three practicals which we have found useful as an introduction to fish embryology. The first consists of a straightforward examination of the gonads of goldfish, the second deals in some detail with the development of a typical egg laying killifish, and the third draws attention to viviparity in fish, using the guppy as an example. A few additional notes are added on material with which we have had little or no experience but which we feel deserves a mention, either because it has been used successfully in the past (trout fertilization), or because it can form the basis of more ambitious practicals.

Goldfish

Despite their popularity as aquarium fish and the accumulated wisdom of countless fish breeders, goldfish are not easy to breed in the laboratory. The favoured methods of breeding involve the use of large outdoor ponds (Hervey and Hems, 1968). The difficulty associated with breeding, and the seasonal nature of the reproductive cycle, severely limits the use of goldfish for developmental studies. For teaching purposes, however, young, mature, goldfish are especially useful for demonstrating in living material, the main feature of teleost gametogenesis. We include a short account of this below.

Examination of the gonads

Fish, about 4 to 5 cm long, are killed by adding a small amount of MS222 to the water containing them. The abdominal cavity is then opened on one side by first making a longitudinal incision along the ventral mid-line, followed by two cuts dorsal-wards, one behind the fins and one in front of the anus. This produces a flap which can be pinned back to expose the body cavity. The general disposition of the viscera is shown in Fig. 5.1. Immediately the body cavity is opened the contents should be kept moist with saline. The gonads lie below the air bladder and above the gut. They are usually quite distinct, the testes being milky white and the ovaries an orange-yellow colour.

When they have been located, the gonads can be easily removed by carefully freeing them from the surrounding viscera. This is accomplished by working backwards towards the vas deferens or oviduct, which must be cut through to release the gonads. Small pieces of tissue are removed from the main mass, placed in a drop of saline on a microscope slide, and mounted gently beneath a cover slip. The slightly squashed

FISH

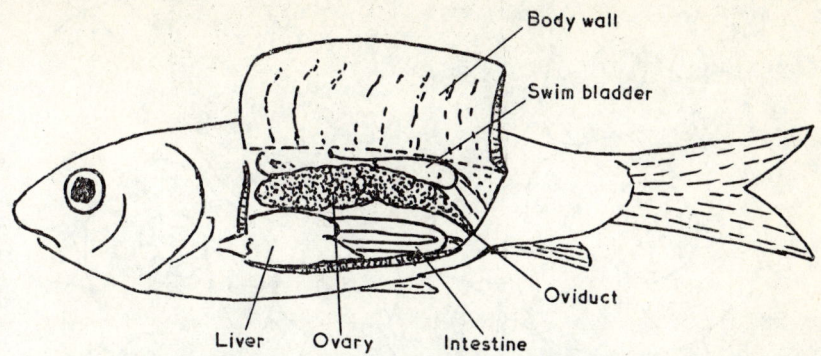

Fig. 5.1. Body wall of goldfish opened on the left side to show the position of the ovary.

tissue forms an excellent microscopic preparation in which various stages of oogenesis or spermatogenesis can be identified. Particularly noteworthy are the cytoplasmic inclusions and the numerous nucleoli of the small and medium sized oocytes (Fig. 5.2).

Killifish

Killifish are egg laying members of the family *Cyprinodontidae* (tooth carps) which are widely distributed throughout the tropical and warmer temperate parts of the world. There are over a hundred recorded species, most of them only a few centimetres in length, and many of them well known to aquarists because of their striking colouration. For practical studies of teleost development these fish are especially useful, and it is of particular value that this potentially rich source of material is well documented (Sterba, 1966; Scheel, 1969). Many species survive well under normal aquarium conditions and can be relied upon to produce a continuous supply of eggs throughout the year. Both the egg and its chorion are transparent, enabling the observation of development from fertilization to hatching. This group also contains a large number of species of 'Annual Fish' which live in waters that normally disappear

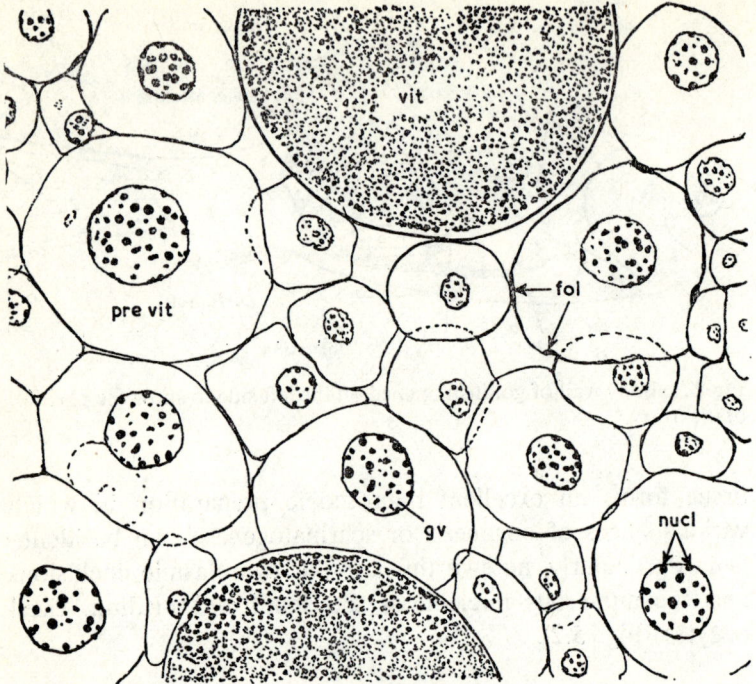

Fig. 5.2. *Drawing of unstained preparation of young goldfish ovary. Abbreviations*: fol – follicle cells; gv – germinal vesicle; nucl – nucleoli; pre vit – pre-vitellogenic oocyte; vit – vitellogenic oocyte.

during the dry season. These drought conditions kill the entire population of active fish and only the eggs buried in the ground survive. When the rains return the embryos hatch and repopulate the temporary ponds. This adaptability is related to the ability of the embryos to suspend their development at particular stages. The nature of such diapauses in a vertebrate embryo is of particular interest to embryologists. The first diapause involves a dispersal, followed by a re-aggregation of the blastoderm cells; the second diapause involves a suspension of organogenesis, and the third, a kind of suspended animation prior to hatching. Wourms (1967) gives a recent and detailed account of these fish.

Killifish vary in their egg laying habits and for practical

purposes it is desirable to choose a species which will lay its eggs on plants or artificial nylon mops suspended in the breeding tank. In considering a suitable laboratory stock other factors must also be borne in mind. The species should be fairly hardy, should not require carefully controlled temperature conditions and should not take too long to develop. Fortunately a fairly rational choice can be made by consulting the popular literature on the subject, (e.g. Turner and Pafenyk list the requirements of nearly a hundred species). Three species which we have found to be both reliable and useful are *Pachypanchax playfairi*, *Aphyosemion scheeli* and *A. australe*, and the following account applies to these fish.

General care

The fish can be kept either in pairs, or in larger numbers in suitably sized tanks, allowing about two litres of water for each fish. The water should be kept at about 20 to 22°C and be soft and slightly acid. Rain water containing a small amount of peat is suitable. About a third of the water in the tank should be replaced every two weeks. The fish are fed on *Tubifex*, white worm, and an occasional meal of a proprietary food, such as 'Tetramin'.

Egg collection and rearing

The eggs are laid on nylon mops or weed floating in the tank. The mops and weeds should be examined regularly and the eggs *gently* removed by using a pair of fine forceps. From time to time the mops are sterilized before placing them back into the tank. When the eggs have been collected, they should be placed in a petri dish containing some of the tank water which has been filtered to remove debris and larger micro-organisms. Each egg should be carefully examined under a dissecting microscope, staged, (see below) and any damaged or dead eggs thrown away. It is important to clean the surface of the chorion at this stage by picking away the fluffy outer coat. If the chorion is cleaned in this way the chances of fungal attack are greatly reduced and a high percentage of the eggs will survive. If kept at 20 to 22°C the embryos will develop into young fish which will hatch within about three weeks. The fry should be fed immediately on newly hatched brine shrimps (see p. 76) or microworms. After about six weeks they can be introduced to the adult diet of *Tubifex* or white worms. The fish will mature in six months to a year.

Development of Aphyosemion scheeli
The following account of the development of *Aphyosemion scheeli* is based on a description of the changes in external appearance of the embryo which takes place during the first four days at a temperature of about 20°C. The development of this fish closely resembles that of *Fundulus heteroclitus* as described by Oppenheimer (1936) and Armstrong and Child (1965). For the purposes of comparison, we have related the description of *A. scheeli* to the stage numbers of Armstrong and Child for *Fundulus,* but in this connection would emphasize that there is no common stage numbering for teleost embryos.

About one hour after fertilization a hump of clear cytoplasm accumulates at the animal pole to form the first cell (stage 2). During the first hour a vertical cleavage divides the first cell into two (stage 3) and during the next hour a second vertical cleavage, at right angles to the first, forms four cells (stage 4). Further cleavages at about half hourly intervals lead first to eight cells (stage 5) and then to 16 cells (stage 6). Synchronous cleavages continue for a while (until the eighth cleavage) and result in the formation of a blastula. At first the blastoderm is raised from the surface of the egg (stage 9) but soon flattens so that it more or less conforms with the general curvature of the egg (stage 13).

The flattening of the blastoderm occurs about 12 hours after fertilization and marks the end of the passive cleavage of the egg cytoplasm. This is followed by an active period of co-ordinated cell movements, which result in the formation of the gastrula and spreading of the blastoderm over the surface of the yolk. In the killifish the expanding blastoderm gives the impression of being under tension as its expanding edge appears to pinch the yolk. The internal pressure created by this movement pushes the lipid droplets, inside the egg, towards the vegetal pole. About 24 hours after fertilization the blastoderm covers about a third of the surface of the yolk (stage 15), and six hours later about two thirds of the surface is covered. The blastoderm completely covers the surface of the yolk after about 40 hours, and it appears quite featureless. Indeed, the cell layers are so thin and uniform at this stage that the embryos can sometimes be mistaken for an unfertilized egg. A few hours later, a streak of tissue can be discerned on the surface of the blastoderm, marking the position of the future embryo.

The next phase involves the elaboration of the main embryonic axis and is dominated by the formation of the head and somites. Two to three somites can be seen some 60 hours after fertilization (stage 21). During the fourth day the number of somites increases rapidly, reaching 20 after about 80 hours, and 30 after about 90 hours. Over this period the heart begins to beat and melanocytes begin to appear, first on the extra-embryonic blastoderm, and then on the embryo itself (stage 23-24). At the same time the head becomes differentiated into distinct regions of fore, mid and hind brain, and it also develops prominent optic cups. Towards the end of the fourth day the beginnings of a vigorous vitelline circulation can be seen, the lens of the eye is obvious and the embryo has a definite tail which undergoes spontaneous muscular twitches (stage 25). The early development of *A. scheeli* is illustrated in Fig. 5.3.

Guppies

The guppy, *Lebistes reticulatus* is a popular aquarium fish and is readily available at most pet shops. It is one of a large number of species of live-bearing tooth carps, and under natural conditions occurs in very large numbers, hence the popular name, 'Millions Fish'. The two sexes are quite distinct, the males

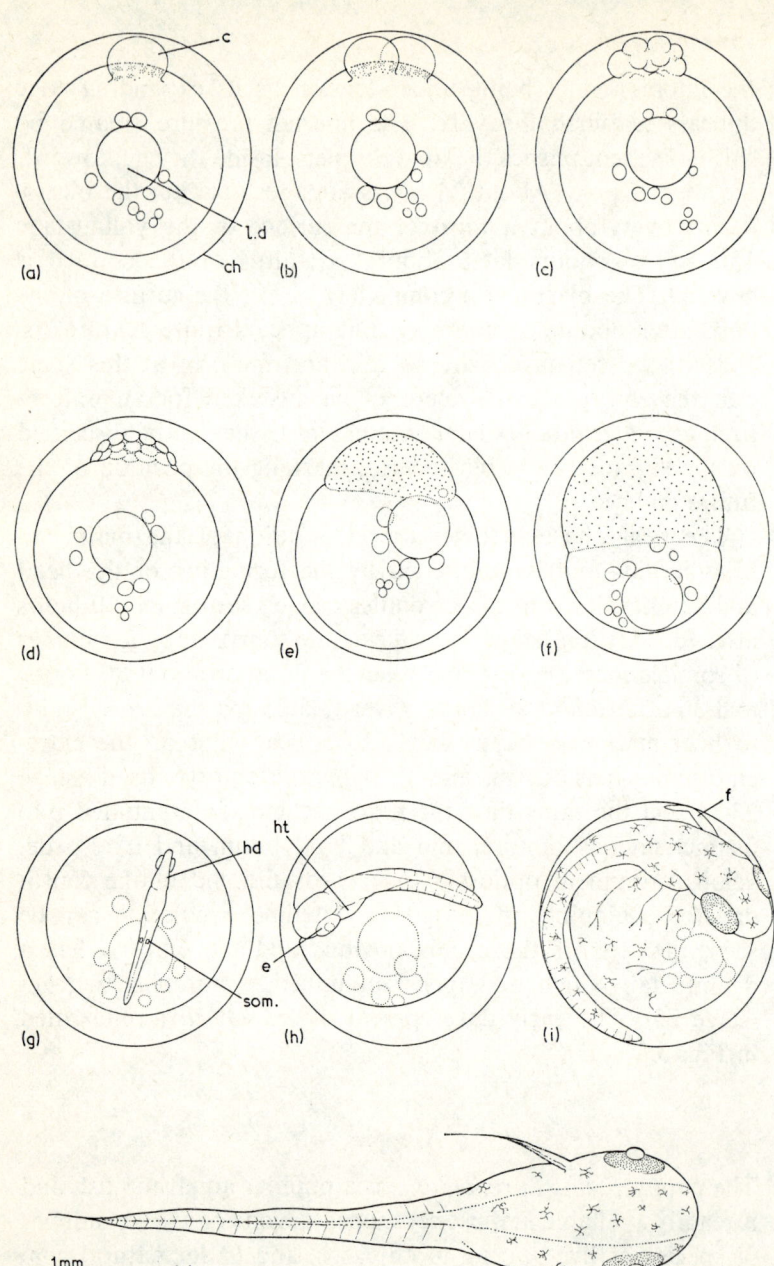

FISH

being more colourful and smaller (about 3 cm) than the larger females (about 5 cm). Guppies are hardy, peaceful and prolific; if they are fed well and kept at about 22 to 24°C they will provide a reliable source of teleost embryos throughout the year. Large numbers of them can be kept in a single tank. The fish are viviparous, with a breeding cycle that lasts about thirty days. After fertilization, the eggs which are inside the ovarian cavity take about three weeks to develop into young fish. As soon as the fry are released from the mother, a new batch of eggs is produced by the ovary, and within ten days, provided sperm is available, a new brood will start to develop. For a general account of the reproductive cycle in *Lebistes* see Rosenthal (1952).

To obtain embryos, selected guppies are killed with MS222 (about 500 μg/ml) and the abdominal cavity opened on one side in a similar manner to that described for the goldfish (see p. 82). As soon as the cavity has been opened the exposed organs should be kept moist with saline. The prominent, yellowish, ovary is distinct and should be examined *in situ* before it is removed. If the embryos are well advanced their pigmented eyes will be visible through the wall of the ovary. Once the ovary has been located, it can be removed and opened in order to examine the embryos in more detail. The average brood size is normally about 12, but 20 embryos are not uncommon, and as many as 70 have been recorded (Purser, 1938). A dozen or so female guppies, selected at random, will provide a large class of students with embryos at all stages of development. The developmental stages are very similar to those described for killifish. However, the guppy egg contains

---

Fig. 5.3. *Stages in the development of Aphyosemion scheeli.* (a) 1 cell (1 h). (b) 2 cells (2 h). (c) 8 cells (3 h). (d) About 32 cells (4 h). (e) Flattening of blastoderm (12 h). (f) Blastoderm covers about two-thirds of yolk surface (30 h). (g) First somite (60 h). (h) Approximately 20 somites (80 h). (i) Pectoral fin movement (8 days). (j) Newly hatched fish (2 weeks).

The times given in parentheses are approximate and correspond to development at 20 – 21 °C. *Abbreviations*: c – first cell; ch – chorion; e – eye; f – pectoral fin; hd – head; ht – heart; l.d – lipid droplets; som – first somite.

more yolk and thus is more opaque than that of the killifish; consequently the earlier stages of development, unlike the later stages, are more difficult to see.

Apart from their embryological interest, guppies can also be used for studies of the reproductive cycle and associated behavioural changes. For instance, when the male guppy matures the anal fin is transformed into a tube-like structure, the gonopodium, which is used for copulation. The formation of this organ is accompanied by other changes, involving the forward movement of the anus and ventral fin and shortening of the body cavity. Such morphological changes, which are under hormonal control, have been likened to metamorphosis, and the system can be exploited experimentally. Thus it has been demonstrated that testosterone propionate induces the above changes in females, and that their development in the male is suppressed by female hormones (Von Euler and Heller 1963). Another interesting aspect concerns the behaviour of the female. In order to mate with the male, the female needs to be at an angle of about 60° to the horizontal, and it can be shown that this angled posture is normally dependent on the oestrogen level. Thus the angle at which the female fish swims can be used as an index of the state of her reproductive cycle (Bretschneider and De Wit Duyvene 1947), see Fig. 5.4.

Artificial Fertilization

Although the artificial fertilization of fish eggs is a practical possibility, it is limited by the seasonal nature of the breeding cycle. Thus in the examples considered below, suitable trout can only be obtained during the winter months, and goldfish only during the summer. In the case of trout, eggs are obtained by grasping a ripe female in a damp cloth in both hands, and *gently* squeezing the abdomen to force the eggs through the cloaca. The eggs are allowed to fall in a dry condition into shallow dishes or trays. Concentrated sperm (called milt), is obtained by treating a male in a similar way, and this is then added to the eggs. The eggs and sperm are mixed thoroughly

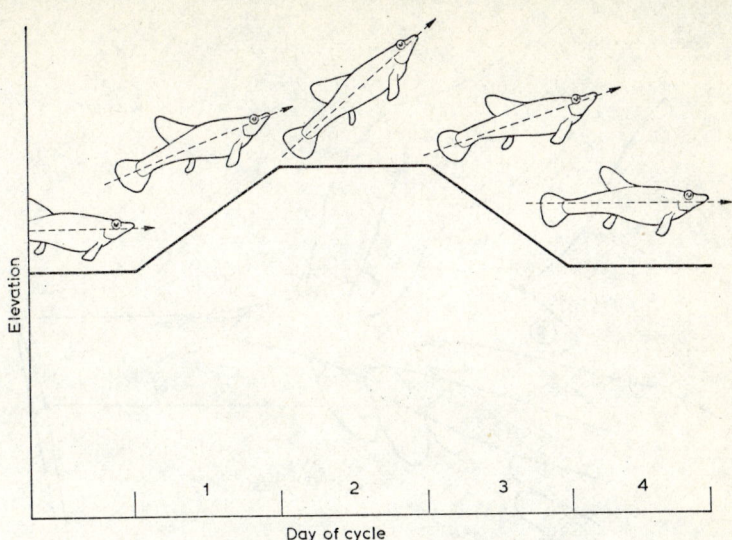

Fig. 5.4. Correlation between angle of swim and reproductive cycle in *Lebistes reticulatum*. (Modified from Bretschneider and De Wit Duyvene, 1947).

and after a few minutes covered with water to a depth of about 5 cm. After an hour the water is removed and replaced with clean water. The best results are obtained if the eggs are left undisturbed in a gentle stream of running water. For this purpose it is best to use trays or dishes with a perforated base. Eggs which die will quickly go opaque and they should be removed as soon as possible. (For further details, see New, 1966).

A similar technique can be used for goldfish (Evans, 1960). The method of stripping these fish is illustrated in Fig. 5.5. In this case it should be emphasized that gentle pressure on the abdomen and not squeezing, is all that is required. Provided really ripe fish are used, artificial fertilization is easy to carry out, produces good results, and has obvious advantages. For instance, an exact timing of development is possible and very large numbers of eggs (up to 2000 in the case of the goldfish) can be obtained. However, it must be stressed that the

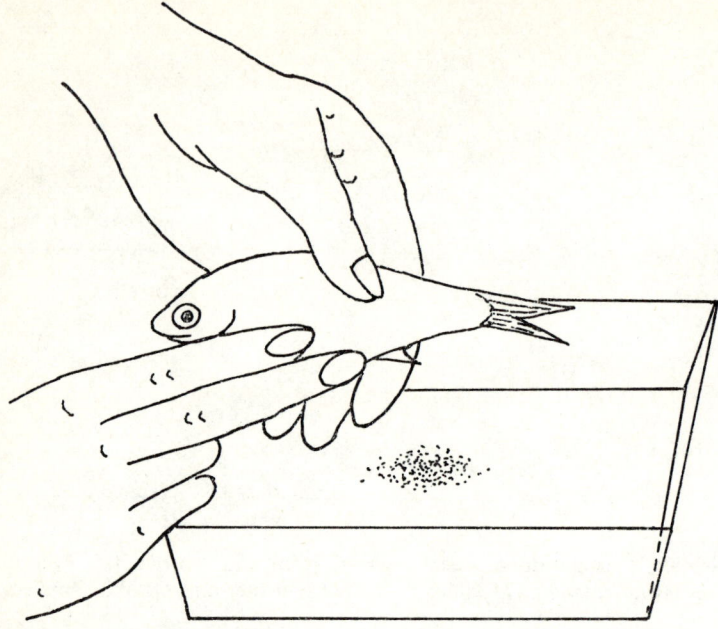

Fig. 5.5 Method of obtaining eggs from a mature goldfish. (Redrawn from Evans, 1960).

seasonal limitation is a severe handicap if one wishes to use the above procedures for general teaching purposes.

Experimental Work on Fish Embryos

Experimental work with fish embryos is technically more difficult than it is with the embryos of amphibia and birds. There are several reasons for this. The eggs are relatively small, they are susceptible to injury and fungal attack, and it is often difficult to obtain large numbers of embryos at a particular stage and at a convenient time. These restrictions do not apply to the other vertebrate embryos dealt with in this book. Therefore, although our experience is limited in this matter, we feel that experimental work using fish embryos is more suited to individual projects than it is to practical classes involving more than a few students. For this reason we shall only draw attention to a few techniques which might prove useful for individual practical work. The reader is referred to Rugh (1962) for a more comprehensive account of the kind of work which is possible with fish embryos.

Decapsulation

The capsule (or chorion) which surrounds the egg, is usually very tough. The method required for removing it seems to depend on the species. In many cases it is possible to remove the capsule enzymatically. For instance pronase can be used to attack the chorion of killifish (Smithberg, 1966), and trypsin will remove that of the stone loach. In others, however, the chorion is very resistant to enzyme attack, and in such cases the chorion has to be removed surgically with iridectomy scissors and fine forceps (e.g. *Fundulus,* Trinkaus, 1966). Once decapsulated the embryos can be cultured in sterile full strength Holtfreter or Steinberg's medium (p 121). In the case of experimental work with fish embryos the sterility of the medium is very important and the addition of a small amount of sulphadiazine (about 100 μg/ml) is recommended.

Explants of the blastoderm

After dechorionation it is possible to remove the blastoderms of embryos at an early stage of development and to culture either the entire blastoderms, or fragments of them, in standard media such as twice normal strength Holtfreter or Steinberg solution. The blastoderm is removed, under sterile conditions, by carefully cutting through the edge of the blastoderm to free it from the remainder of the egg. The isolated blastoderms can be used for studying the effect of changes in the medium on growth and differentiation (Trinkaus and Drake, 1956). If large numbers of embryos are required for biochemical work, then it is possible, in some cases, to separate the blastoderm from the yolk by centrifuging the de-chorionated embryos in a sucrose density gradient (Kostomarova, 1967).

Experimental analysis of development
The experimental analysis of teleost development has, to a large extent, been centred on problems concerning the regulative capacity of the early blastoderm and the significance of the co-ordinated cell movements which take place as the blastoderm covers the surface of the egg and organizes itself to form an embryo. The regulative capacity of the early blastoderm can be investigated by the destruction, or removal, of one or more cells during the early cleavage stages. The cells are destroyed by pricking them with a fine needle which can either be passed directly through the chorion or through a small window cut in the chorion. Alternatively, a fine mouth-controlled pipette can be used to remove one or more cells from the blastoderm. Using these methods (in *Fundulus*) Nicholas and Oppenheimer (1942) were able to show that the initial blastomeres (up to the 16 cell stage) were totipotent, and that even if large parts of the blastoderm were removed a normal embryo could be obtained. The morphogenetic movements of teleosts can be studied by vital staining, simple operative procedures and time-lapse photography (Trinkaus, 1969).

REFERENCES

Armstrong, P. B. and Child, J. S. (1965), 'Stages in the norma development of *Fundulus heteroclitus*', *Biol. Bull.*, **128**, 143–168
Bretschneider, L. H. and De Wit Duyvene, J. J. (1947), *Sexual Endocrinology of Non-mammalian Vertebrates*, Elsevier: New York, Amsterdam.
Cohen, J. (1967), *Living Embryos*, Pergamon Press: London.
Evans, A. (1960), *The Goldfish*, Foyles, London.
Hervey, G. F. and Hems, J. (1968), *The Goldfish*, Faber and Faber, London.
Kostomarova, A. A. (1967), 'The differentiation capacity of isolated loach (*Misgurnus fossilis*) blastoderm', *J. Embryol. exp. Morph.*, **32**, 407–430.
New, D. A. T. (1966), *The Culture of Vertebrate Embryos*, Logos: Academic Press: London.

Nicholas, J. S. and Oppenheimer, J. M. (1942), 'Regulation and reconstitution in *Fundulus*' *J. exp. Zool.*, **90**, 127–158.

Oppenheimer, J. M. (1936), 'The Normal Stages of *Fundulus heteroclitus*', *Anat. Rec.*, **68**, 1–15.

Purser, G. L. (1938), 'Reproduction in *Lebistes reticulatus*', *Quart. J. Micros. Sci.*, **81**, 151–157.

Rosenthal, H. L. (1952), 'Observations on reproduction in the poeciliid *Lebistes reticulatus* (Peters). *Biol. Bull.* **102**, 30–38.

Rugh, R. (1962), *Experimental Embryology*, Burgess Publishing Co: Minneapolis.

Scheel, J. (1969), *Rivulins of the Old World*, Publ. T.F.H.: Inc.: Reigate/New Jersey.

Smithberg, M. (1966), 'An enzymatic procedure for dechorionating the fish embryo, *Oryzias latipes*', *Anat. Rec.*, **154**, 823–829.

Sterba, G. (1966), *Freshwater Fishes of the World*, Vista Books: London.

Trinkaus, J. P. (1966), 'Fundulus', In: *Methods of Developmental Biology*, Ed. F. H. Wilt and N. K. Wessells, Crowell Co., New York.

Trinkaus, J. (1969), *Cells into Organs – The Forces that Shape the Embryo*, Prentice Hall, New Jersey.

Trinkaus, J. and Drake, J. W. (1956), 'Exogenous control of morphogenesis in isolated *Fundulus* blastoderms by nutrient chemical factors', *J. exp. Zool.*, **132**, 311–348.

Turner, B. J. and Pafenyk, J. W. *Enjoy Your Killifish*, The Pet Library Ltd., New York.

Von Euler, U. S. and Heller, H. (1963), *Comparative Endocrinology*, Volume I, Academic Press, New York and London.

Wourms, J. P. (1967), 'Annual Fishes', In: *Methods of Developmental Biology* Ed. F. H. Wilts and N. K. Wessells. Crowell Co, New York.

6 Amphibia

Our understanding of the fundamental mechanisms of animal development in general, and of vertebrate development in particular, owes much to work on amphibian embryos. This work spans the whole of the period of experimental embryology in the modern sense, starting with the establishment of fundamental operative techniques at the turn of the century and extending to contemporary studies of nuclear transplantation and biochemical analysis. The usefulness of amphibians is not confined to their embryos; the metamorphosis of their larvae and their capacity for regeneration have also contributed much to developmental studies. This emphasis on amphibian development is reflected in the special attention given to amphibia in all elementary texts dealing with embryology. This is true not only of texts of a general kind but also of the majority of practical manuals. In view of the wealth of material already available it is perhaps presumptuous to add yet more to the countless descriptions and practical suggestions already made. Yet we do so for two reasons. The first is the weaker one, namely, for the sake of completeness of the practical approach to development. Any account with no more than a mention of these embryos would be deservedly considered unbalanced and peculiarly biased. The second, and more positive reason, is the belief that to record our own practical approach will prove useful not only to those who have never tried experiments on amphibian embryos but also to those, who using other methods, have tried and failed. This is not to say that other methods are wrong or that success is guaranteed by those described here; care in preparation, patience in observing and handling the embryos, and a previous attempt to understand the developmental process is the way to success, whatever guidelines have been laid down for the handling of these or any other embryos.

Many of the experiments and the normal tables of development described in the literature concern anurans and urodeles which have been obtained from natural populations, and there is no doubt that species belonging to *Rana* and *Triturus* will continue to be used for research purposes, proving as useful in the future as they have done in the past. However, for teaching purposes such animals are not really to be recommended; they have limited breeding seasons, there is the additional effort involved in their location and capture, and the knowledge that each haul involves a further depletion of a natural population. With this in mind we have limited the present account to experiments and observations on *Xenopus* and axolotls which, although they can be obtained from natural populations, have the considerable advantage that they can be easily maintained and bred in the laboratory. Both species have been used extensively in developmental studies, and practical work on their embryos and larvae can be carried out against a considerable background of established knowledge.

In the following sections both species are dealt with in some detail, although in view of its greater usefulness at an elementary level more attention is paid to *Xenopus*. In both cases accounts are given, first of the care and maintenance of stocks, the breeding cycle and the production of eggs, and of the embryonic development. This is followed by details of experiments which we consider feasible for class practicals, followed by a mention of others which might be carried out on an individual basis.

Xenopus

The South African clawed toad, *Xenopus laevis* is a fairly hardy and completely aquatic anuran which can be kept under fairly simple conditions. It is most convenient to keep the animals in pairs, each consisting of a mature male and female, in glass tanks with angle iron or plastic frames. A tank measuring about 20 × 20 × 45 cm is a useful size and it should be not more than one third full of water. *Xenopus* are strong swimmers and very agile animals and will certainly escape from a tank of this size if more than the suggested amount of water is used. In any case, a heavy cover, such as a piece of plate glass, should be placed over the tank. The temperature at which the animals are kept does not seem to be very important although they probably do best at a temperature of around 20°C. This condition is easily obtained in most rooms by using a domestic type electric convection heater which is thermostatically controlled. However, we have kept animals for long periods (a year or more) in an unheated aquarium room where the temperature fluctuated from below 10 to above 20°C without noticeable effects on the animals' general health or breeding capacity. The most important factors in the care of these animals is a regular supply of food and regular cleaning. There is no need to simulate near tropical conditions. As far as feeding is concerned our own practice is to feed the animals two or three times a week on beef heart or liver chopped into small cubes of about 5 mm. About a dozen pieces for each animal, supplemented by an occasional meal of *Tubifex*, seems to be sufficient. Most importantly the water in the tanks should be changed at least once a week and the tanks thoroughly cleaned at the same time. Under good conditions disease is extremely rare in these animals. If it does occur it is best to kill the animals at once rather than to prolong their suffering by attempting a cure.

Induced spawning in Xenopus
A major advantage of *Xenopus* over other amphibians is that fertilized eggs may be obtained in large numbers throughout the year from mature animals by the injection of small amounts of chorionic gonadotrophin. Successful ovulations may be obtained by the following procedure. The animals are removed from their storage tanks and placed in a round glass trough, about 30 cm in diameter and about 12 cm deep (known to chemists as a pneumatic trough) which is half filled with water.

A heavy piece of plate glass is placed over the trough to prevent the escape of the animals. The trough should be placed where the animals are unlikely to be disturbed and where there is a fairly even temperature of about 20°C (i.e. normal laboratory temperature). Temperature control of the individual containers is not necessary. Once the pairs have been selected they should be left in their troughs for about 24 hours, without feeding, before they are injected. If a practical class is at stake more than one pair must be injected to ensure a supply of eggs. We recommend that a minimum of three pairs, preferably four, be used on each occasion. After the period without food, and immediately before the injections, the water in the troughs is replaced. When eggs are required each animal is given two injections, spaced at an interval of 24 hours. On the first day the female receives 100 units of gonadotrophin and the male 50 units. On the second day the dose is doubled, i.e. the female receives 200 units and the male 100. These amounts are for normal sized animals; for very large animals the dose should be increased by about 20 per cent. It is best to carry out the injections during the late afternoon or early evening; this will usually give eggs at an early stage of development on the morning of the third day, with the animals remaining clasped during the day and producing more fertile eggs.

The hormone should be dissolved in clean distilled water which need not be sterile. The amount of water used should be calculated to give an injection volume of about 0.5 ml for each animal at each injection. A 2.0 ml graduated syringe is a useful size. A fairly large needle is required; we normally use a 21G needle, $1\frac{1}{2}$in long (40 mm 8/10). The object of the injection is to introduce the hormone into the dorsal lymph space. Although this can be done single handed it is best to have assistance, so that one person can hold the toad and the other concentrate on the injection. The method is illustrated in Plate 6.1. The toad is held in the left hand with the thumb and forefinger crooked around the left hind leg with the rest of the hand passing round the body. The object of this is to expose the upper part of the leg and the adjacent area of skin which covers the dorsal lymph space. The animal is held with the ventral side downwards on a damp cloth and the end of the left leg held at its lower end with the free hand, using part of the cloth to secure the grip. The palm of the left hand covers the upper part of the body and the head of the animal and prevents its escape. The needle is inserted into the exposed leg about mid-way along the thigh into a loose channel formed by the muscles. It is then pushed gently beneath the surface of the skin towards, and then into, the dorsal lymph space. Throughout the procedure the progress of the needle should be observed and the location of the tip known. The needle should penetrate about 1 cm into the lymph space, should always be superficial and should be kept well clear of the spinal column. Once in position the injection is made smoothly and the needle withdrawn slowly, using a finger tip to prevent fluid loss as it is finally removed.

The injection is likely to have been successful if the cloaca of the female becomes red and slightly swollen after the priming dose. If both male and female respond they will go into amplexus some hours after the second dose and the female will begin to lay eggs on the morning of the third day. A mature female can produce well over a thousand eggs at a single laying, but five or six hundred is a more likely score. Although animals can be injected again after six weeks, it is best to leave them for about three months before a further attempt is made to produce more eggs. After three or four years the animals become less responsive to hormone and should be replaced.

It is advisable to collect the eggs as soon as possible after they are laid as the parents will occasionally eat them. The newly laid eggs stick to the glass and must be detached gently with fine forceps. Rough handling at this, the cleavage stage, will damage the eggs and cause unnecessary loss. Once loose, the eggs can be removed with a wide mouthed pipette. The jelly coats of the eggs are quite transparent and the developing embryos are easily observed through them with the aid of a dissecting microscope. A black background, using top illumination, gives an excellent view of the cleavage and subsequent stages. Eggs which have not been fertilized, or which have been damaged, will swell and become white. They are easily recognized and should be removed as soon as possible because their presence in large numbers will foul the water and ruin the rest of the batch.

The development of Xenopus laevis
The embryonic development of *Xenopus* is fairly typical of amphibians generally, but the details of the gastrulation process do not appear to be quite the same as in other amphibia (Deuchar, 1966). A close study of the external features of the embryo as it develops will, however, provide a good example of amphibian development. A stage numbering system is commonly used to describe the stages of development, and in amphibians the early stages are usually referred to by corresponding numbers when one stage is compared with another. Thus Stage 10 indicates the beginning of gastrulation and Stages 12 and 13 represent its end. After neurulation however, the stage numbers assigned to the embryos of different species are not always comparable, and it becomes especially difficult to compare the post-neurula stages of anurans and urodeles. For both *Xenopus* and the axolotl, we have concentrated on those early developmental stages which the embryos have in common in order to lay emphasis on the early features of amphibian development. We have dealt with the later embyronic and larval stages less fully with experimental work rather than descriptive detail in mind. The following summary is based on the stages of Nieuwkoop and Faber (1967); particular attention is paid to those features which indicate important changes in the developing embryo.

Recently fertilized egg (stage 1). Note that the animal surface is pigmented and the lower vegetal surface is not. The concentration of the yolk in the vegetal half orientates the embryo so that it remains with the animal half uppermost until the end of gastrulation.

Two cells (stage 2). About two hours after it is laid the egg begins to divide. The cleavage furrow appears first in the animal half. It deepens and extends to the vegetal side. At the end of the third hour cleavage is complete.

Four cells (stage 3). The second cleavage furrow appears at right angles to the first and takes about half an hour to completely divide the egg into four cells.

Eight cells (stage 4). The third cleavage is horizontal and divides the egg into eight; four animal cells and four larger vegetal cells. This cleavage normally takes 10 to 15 minutes.

Cleavage to 'medium cell blastula' (stages 5 to 8). During this period cleavages begin to occur more rapidly (about one in every 10 minutes) and the cells become smaller and smaller. The presence of yolk retards the cell division in the vegetal half and leads to a marked difference in cell size between the two halves of the embryo. Such cleavage leads to the gradual formation of a space (the blastocoele) inside the embryo. At the 'medium cell blastula' stage, the cells at the animal pole are just discernible under the low power of a dissecting microscope.

Fine cell blastula (stage 9). This is the definitive blastula stage and the individual cells are now rather difficult to see with a dissecting microscope.

Initial gastrula (stage 10). The first sign of gastrulation is the appearance of a small pigmented line just below the equator of the blastula and in the yellowish, vegetal half of the embryo.

Early gastrula (stage 11). The dorsal lip of the blastopore becomes crescent shaped and then horseshoe shaped and eventually forms a complete ring which encloses the yolky cells. When the ring is complete its diameter is about half that of the egg.

AMPHIBIA

Late gastrula (stage 12). The enclosed yolky cells now take the form of a yolk plug which is about a quarter of the diameter of the egg. Gastrulation takes about five hours to complete. Towards the end of the process the yolk plug becomes ovoid in shape (stage $12\frac{1}{2}$) and then slit like, as the neural plate begins to form.

Formation of the neural plate (stages 13 and 14). Gastrulation brings the roof of the archenteron into contact with the overlying ectoderm. As a result of this an interaction takes place between the two cell layers and a neural plate is formed from the ectoderm along the future dorsal surface of the embryo. This is the primary inductive interaction. The plate will be seen as a flattened, almost pear shaped area. Careful observation will show that the area is slightly less pigmented than the surrounding ectoderm and, at the initial stage, more or less pigmented lines will be seen radiating from the yolk plug.

Formation of the neural folds (stages 15 to 17). Folds appear along the edge of the plate and gradually move towards the mid-line. These are the neural folds and their formation transforms the plate into a groove. At the beginning of the formation of the folds a pigmented patch appears at the anterior end of the embryo. This is the adhesive gland.

Formation of the neural tube (stages 18 to 21). The tube is formed by the fusion of the folds, which begin to move towards each other in the mid-line, in the trunk region, (stage 18) first making contact in this region (stage 19). During this period the embryo begins to change shape, its lateral outline becoming convex. The folds gradually become fused along the mid-dorsal line, although the suture remains visible for some time (stage 20). Eye rudiments are just visible, protruding on each side of the anterior part of the head. Further changes of shape occur as the lateral outline becomes at first flat and then concave (stage 21). At this stage the neural tube is completely closed and the embryo begins to lengthen.

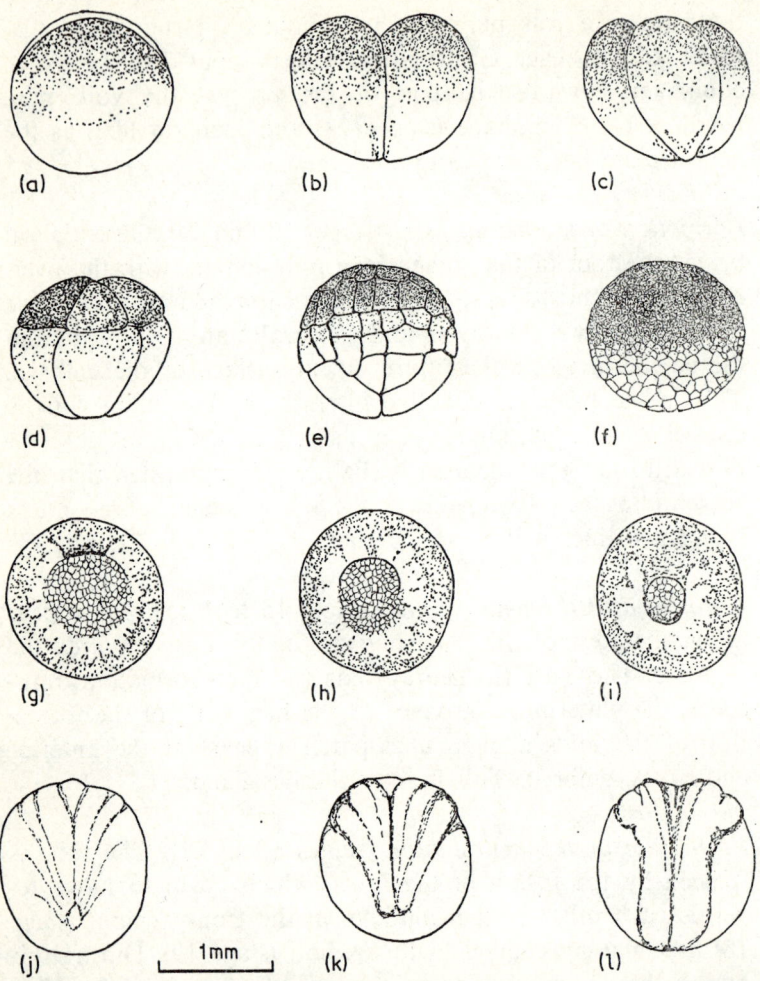

Fig. 6.1. *Stages in the development of Xenopus laevis I.* Stages according to Nieuwkoop and Faber. (a) Recently fertilized egg, viewed from the side. (Stage 1). (b) Two cells, viewed from the side (Stage 2). (c) Four cells, dorso-lateral view (Stage 3). (d) 8 cells, dorso-lateral view (Stage 3). (e) Mid cleavage (Stage 6). (f) Late cleavage (Stage 8). (g – i) Gastrulation, viewed from the animal pole (Stages 10, 11 and 12). (j – l) Formation of Neural plate and folds, posterio-dorsal view (Stages 13, 14 and 15). Selected and redrawn from Nieuwkoop and Faber (1967).

The closure of the neural tube marks the end of an initial period in which single changes can be seen to affect the embryo as a whole. These early stages of development are illustrated in Fig. 6.1. The next period is one in which individual organ rudiments begin to develop. The presence of these rudiments can be detected externally and are used as further criteria for staging the embryos.

Temperature affects the time that it takes an embryo to reach a particular stage. *Xenopus* embryos develop normally over a temperature range of 15 to 30°C, and a knowledge of the rate at which an embryo will approach a given stage within this range is often very useful when planning experiments. For instance, if two different stages are required from the same batch of eggs, then, provided the stages are not too far apart, this can be achieved by using a temperature differential. The graph published by Bebbington and Thompson (1967) gives a rough guide to the effect of temperature on the development of *Xenopus* until hatching. From this it may be seen that at 15°C gastrulation begins at about 30 hours and ends at 45 hours, and that stage 24 is not reached until about 100 hours. At 25°C, development is very much quicker, gastrulation is reached within 10 hours, and stage 24 within 30 hours. By focussing attention on stage criteria, a study of the effect of temperature on development forms an instructive class exercise, and the accumulated data are useful in the planning of future experiments.

It is not our intention to give a detailed and progressive account of the later stages of development. This material is extremely well documented in the normal tables of Nieuwkoop and Faber and no purpose would be served by attempting to summarize all their descriptions here. However, we have selected a few of these stages to indicate the development of the embryo to the feeding larval phase and also as an aid to the identification of the stages referred to in the experiments described later in this chapter.

The stage numbers are those of Nieuwkoop and Faber and

the descriptions take account of their external criteria (Fig. 6.2).

Stage 25 (Fig. 6.2a). At this stage several prominent features on the surface of the embryo mark the development of important organ rudiments. Anteriorly, the optic vesicles bulge out on each side of the head just above the adhesive gland. Immediately behind the developing eyes can be seen the gill area, which is divided by grooves at this stage. A small pigmented pit, between and above the eye area, marks the initial development of the otic vesicle. The back of the embryo is still curved and along the mid dorsal line a fin is beginning to develop.

Stage 28 (Fig. 6.2b). The features mentioned above are now more distinct. In particular the fin becomes more prominent and extends around to the anal region, giving the beginnings of a tail to the embryo. The back of the embryo is now quite straight, emphasizing the small increase in length which has taken place.

Stage 35/36 (Fig. 6.2c). During the third to fourth day, if reared at 20°C, the embryos begin to hatch. Distinctive features at this stage are the presence of melanophores along the back, in addition to those which have previously appeared in the anterior flank region. The eye rudiments are darkly pigmented and the choroid fissure is not quite closed. The tail is about three times as long as it is broad.

State 45 (Fig. 6.2d-e). After four to five days the embryo is obviously a tadpole with a large broad head and a long tail. The eyes and coiled gut are prominent. The tail now possesses a broad median fin both dorsally and ventrally. At this stage the animal, with its embryonic yolk reserves exhausted, begins its ceaseless activity in connection with feeding.

After a lengthy feeding phase, which may last about eight weeks, the tadpoles undergo the drastic changes which charac-

AMPHIBIA

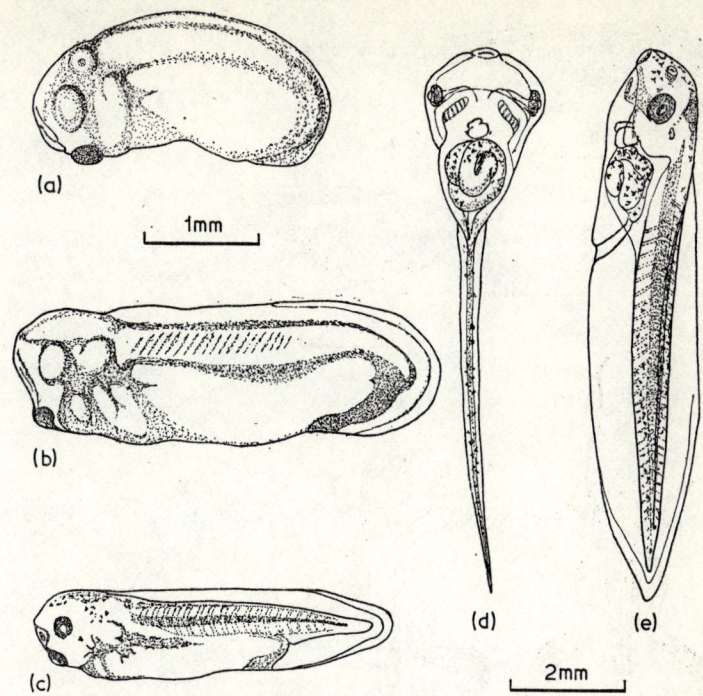

Fig. 6.2. *Stages in the development of Xenopus laevis II.* Staged according to Nieuwkoop and Faber. Descriptions of these stages are given in the text (a) Stage 25. (b) Stage 28. (c) Stage 35/36. (d – e) Stage 45. Selected and redrawn from Nieuwkoop and Faber (1967).

terize metamorphosis. These are illustrated in Fig. 6.3, and are briefly described in connection with a later section. The anatomy and physiology of the feeding larval phase is well worth studying. Many important features of vertebrate head structure can be discerned through the transparent areas of the tadpole head. The main excretory organ, the pronephros, and the heart and associated blood vessels are also easily distinguished. As a demonstration of living vertebrate anatomy, this animal has much to recommend it.

Notes on feeding Xenopus tadpoles
Of the various methods which have been suggested for feeding tadpoles we have found nettle powder and Complan satisfac-

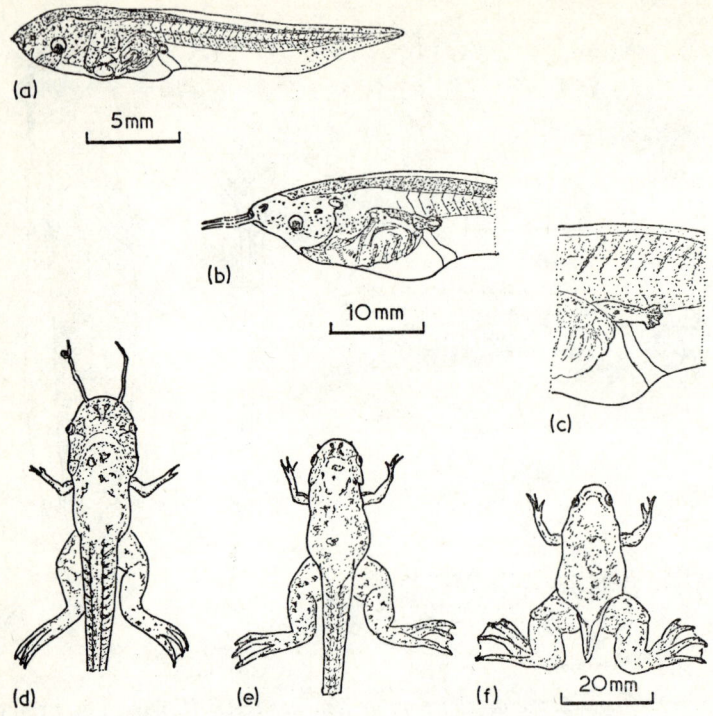

Fig. 6.3. *Stages in the metamorphosis of Xenopus laevis.* Staged according to Nieuwkoop and Faber. (a – c) Prometamorphic tadpoles, stages 50, 53 and 55, showing development of the hind limb. (d – f) Stages 60, 62 and 64, illustrating features of the metamorphic climax. Selected and redrawn from Nieuwkoop and Faber (1967).

tory, and prefer to use a mixture of both. Initially the embryos should be allowed to remain in the water in which they have hatched at a density of several hundred per litre. At stage 45, or thereabouts, they will feed on the debris and micro-organisms which have developed naturally; a sure way to kill most of them is to remove the young larvae to clean water. Once the feeding stage is reached, about one third of the water should be changed and the volume increased if necessary. Food is added by sprinkling the surface of the water with fine nettle powder, or by adding 1 to 2 ml of approximately 1 per cent

Complan suspension. At no stage should the water become cloudy. Dead animals and accumulated rubbish must be removed with a pipette. Feeding (twice a week) and partial change of water (usually not more than once a week) is continued until the animals metamorphose. The growth rate of the tadpoles and the rapidity with which they reach metamorphosis depends on temperature, type and amount of food, and in particular on the density of animals in the water. About ten animals per litre will give a good growth. At the onset of metamorphosis the animals stop feeding and become very inactive. At this stage they need to be placed in shallower water, about 5 cm deep. The young toadlets are fed on *Tubifex* and after about three months gradually transferred to the adult diet of beef heart or liver (see p. 98). Under normal conditions the animals become mature within a year to 18 months.

Axolotls

Axolotls (*Ambystoma mexicanum*) are not difficult to keep provided a few simple criteria are kept constantly in mind. To quote Humphrey (1962) 'axolotls do not require running or aerated water, provided the containers are of suitable size and are kept clean'. We would strongly emphasize the word clean and add that they should be kept cool, and wellfed, but not overfed. Mature animals are conveniently kept, either singly or in mated pairs, in glass aquarium tanks measuring approximately 25 cm wide by 25 cm deep and 40 cm long. These measurements should be regarded as minimum requirements. About half a dozen pieces of broken tile or brick are placed on the bottom of the tanks, which should be about three-quarters full of chlorine-free tap water. There is really no need to cover the bottom of the tank with sand or gravel. It is, in fact, convenient if fairly large areas of the bottom of the tank are kept clear to simplify the processes of feeding and cleaning. Three feeds of small pieces of beef heart a week form a basic diet for the adult animals but feeds of beef liver and well washed *Tubifex*, about once a fortnight, are recommended to keep the animals in good condition. The animals will usually pick up food from the bottom of the tank but if they are reluctant to do so, they can be encouraged to feed by allowing them to snap at small pieces of meat held in blunt forceps. It is essential to replace the water in the tanks once a week, preferably twice. The tanks should be cleaned thoroughly once a fortnight. Every effort should be made to keep the aquarium fairly cool, about 18°C or below. Such conditions are not too difficult to maintain in the UK in an unheated room, but it is very useful to have an air cooling device during the summer months. The chief worry with axolotls is fungal infection. An attack on the gill region is not uncommon, but as this is really a sign of a dirty tank it can be avoided. Slight fungal infections can be cured by adding to the water in the tank a few ml of a dilute solution of methylene blue or malachite green (in each case about 0.1 per cent). Severe infections of fungus and other poorly defined diseases are difficult to treat; attempted cures usually do little but delay the death of the animal. In such cases it is better to kill the animals before a hopelessly moribund condition sets in.

Spawning in axolotls
As fertile matings cannot be induced in axolotls by the injection of hormones, and as artificial fertilization cannot be achieved without killing the animals (see below), the only satisfactory method of obtaining fertilized eggs is to allow the animals to mate during the breeding season. This season lasts from November to May, although eggs are occasionally produced earlier than this (October) and sometimes later (June). The peak period for eggs extends over the winter months of December, January, and February.

If eggs are not required at any particular time, it is best to leave the animals in mated pairs and to keep a close check on the tanks. About 24 hours warning of a successful mating is given by the deposition of several spermatophores by the male. These take the form of conical masses of jelly about 5 mm high, tipped with a white blob of sperm. They are attached to the glass floor of the tank and to flat pieces of brick and tile lying at the bottom of the tank. If the spermatophores are picked up by the female, eggs will be laid on the following day. As many as five or six hundred eggs will be laid slowly over a period of 48 hours or so. The eggs will be attached to objects in the tank such as tubing, glass rods, and pieces of brick and tile. We have found large fragments of brick and tile very useful as the female tends to lay her eggs along the edges of this material from where they can be easily picked off with fine forceps. This must be done gently as pressure will cause abnormal cleavages and the embryos will be lost. It is a good idea to clear the eggs from the tank towards the end of the first day because, although any disturbance may stop the female laying for an hour or so, a set of uncleaved eggs (stage 1) can be obtained when the female starts up again. Alternatively, the female can be placed in a separate tank after she has started laying. During the breeding season spermatophores are often produced soon after a tank has been cleaned. This is clearly associated with the sudden drop in water temperature which accompanies the process. Most mature females can be relied upon to produce one good batch of eggs during the breeding season; a second, and very occasionally a third batch, containing fewer eggs, may follow the first after an interval of a month or so.

A more exact timing of egg production is obtained by isolating males and females at the beginning of the breeding season and reuniting them when eggs are required. If the water is changed and slightly cooled (a drop of about 5°C) when the animals are brought together the chances of success are increased. Once in the presence of a female a previously isolated male will often become very active and will sometimes deposit spermatophores within an hour of being paired. In this case the vigorous movements of the animals often dislodge the spermatophores, which are then useless. The more placid activities of the permanently mated pairs are more certain to produce the desired result.

The development of axolotls

The initial stages of development are very similar to those of *Xenopus* and are described in the same way. There is, however, a significant difference between the two species in the rate of their embryonic development. For instance, at approximately 20°C it takes an axolotl embryo about 48 hours to reach the end of gastrulation, but only about 20 hours for *Xenopus* to reach the same stage. The large size of axolotl eggs and the slowness with which they develop makes them ideal material for detailed observation of early amphibian development. They are especially good for studying cleavage, gastrulation and the formation of the neural plate and folds. There is no need to decapsulate the embryos completely in order to observe their development but a clearer view is obtained if the soft, diffuse, outer jelly layer is removed. This can be done simply by rolling the jelly, containing the embryo, on a piece of filter paper. The soft jelly adheres to the paper, leaving the embryo in its tough inner capsule. The preponderance of yolk in the vegetal half of the egg makes it difficult to view the vegetal region during the early stages of development. The embryo rotates freely inside its capsule and during cleavage and gastrulation always presents the pigmented animal surface uppermost. No matter how carefully, or quickly, one attempts to turn the embryo over, no more than a glimpse of the surface of the vegetal half is obtained as this heavier half disappears from view. The simplest way to see this hidden side is to observe its reflection in a plane mirror placed immediately below the container (a small petri dish). The reflection can be viewed easily with the aid of a dissecting microscope ($\times 10$) if both mirror and container are tilted at an angle of about 20° to the stage of the microscope, and the light source is focussed on to the embryos in the same direction as the mirror and container (Fig. 6.4).

The developmental sequence for the axolotl, up to the formation of about 12 somites, is summarized below. The stage numbers refer to those of Harrison for the related *Ambystoma punctatum* (Hamburger, 1960). The timing relates to a temperature of about 20°C.

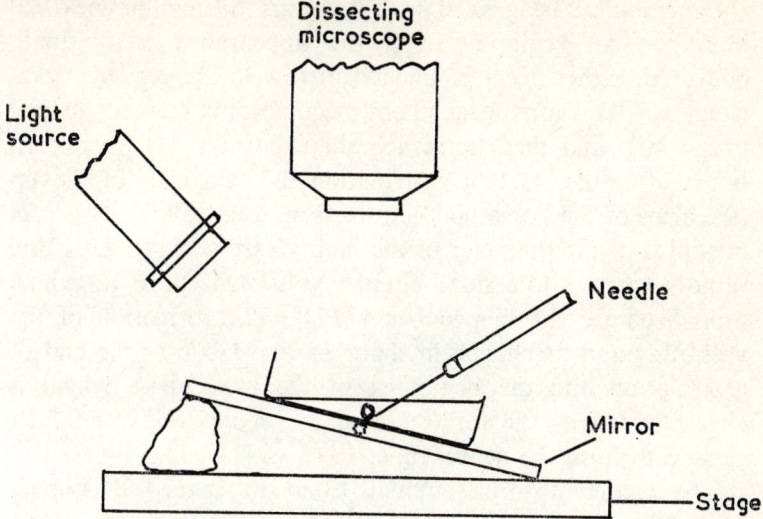

Fig. 6.4 Use of angled mirror to examine vegetal region of amphibian embyro during gastrulation.

The recently fertilized, and consequently matured, egg, possesses a prominent pale area located at the animal pole. This disappears about an hour before the first cleavage, and at this stage the *uncleaved egg* possesses a fairly uniformly pigmented animal surface (stage 1). The first sign of a cleavage furrow appears at the animal pole some six hours after fertilization. The furrow extends from animal to vegetal pole during the course of an hour and leads to the formation of the *two cell stage* (stage 2). A second furrow at right angles to the first, and also taking about an hour to complete, produces the *four cell stage* (stage 3). The *eight cell stage* (stage 4) is formed by a horizontal cleavage, just above the equator, and is reached about ten hours after fertilization. The *16 cell* (stage 5) and approximately *32 cell* (stage 6) stages follow more quickly and further cleavages at about 30 minute intervals lead to *mid cleavage* (stage 7). This is followed by a *late cleavage* stage (stage 8). Cleavage ends with the formation of the *blastula* (stage 9), about 30 hours after laying.

Gastrulation (stages 10 to 12) occurs during the next 20 hours or so, beginning with the appearance of a small, puckered, rather deep, pigmented furrow in the vegetal region (stage 10). This furrow lengthens, becomes first crescent shaped (stage $10\frac{1}{2}$) and then horseshoe shaped (stage 11). About 15 hours after the start of gastrulation the two ends of the lip (the arms of the horseshoe) join to form a complete ring about one-third of the diameter of the embryo (stage $11\frac{1}{2}$). This ring rapidly narrows to a small circular yolk plug (stage 12) which soon becomes slit shaped (stage $12\frac{1}{2}$). The formation of the yolk plug and its change in shape can be taken as the end of gastrulation and the beginning of the next phase which is characterized by the formation of the neural folds and their closure to form the neural tube.

The change from a circular to a slit shaped blastopore passes through stages in which the opening is first shaped like a tear drop and then like a key hole (Fig. 6.5 k–l). The slit of the key hole lies in the direction of the future mid-dorsal line of the embryo and points to the future anterior end. At this stage there is usually some indication of a furrow in the mid-dorsal line, just in front of the closing blastopore. At about this time the first sign of the *neural plate* is indicated by a change in pigmentation and slight flattening of the surface in the region of the closure (stage 13). Soon after this the entire area of the plate becomes demarcated by the incipient neural folds (stage 14). As the plate forms, the embryo begins to flatten from side to side, the furrow in the mid line extends and deepens, and the neural folds begin to rise up on each side of the mid-dorsal line (stage 15). As neuralation continues, the folds become more prominent and move towards each other (stages 16 to 18), touching first in the middle region. The initial contact extends backwards and forwards until the folds meet more or less continuously along the mid-dorsal line (stage 20). The changes which take place during neurulation and during the preceding phases of gastrulation and cleavage are illustrated in Fig. 6.5 and Fig. 6.6 (a–d).

After the closure of the neural folds individual organ rudi-

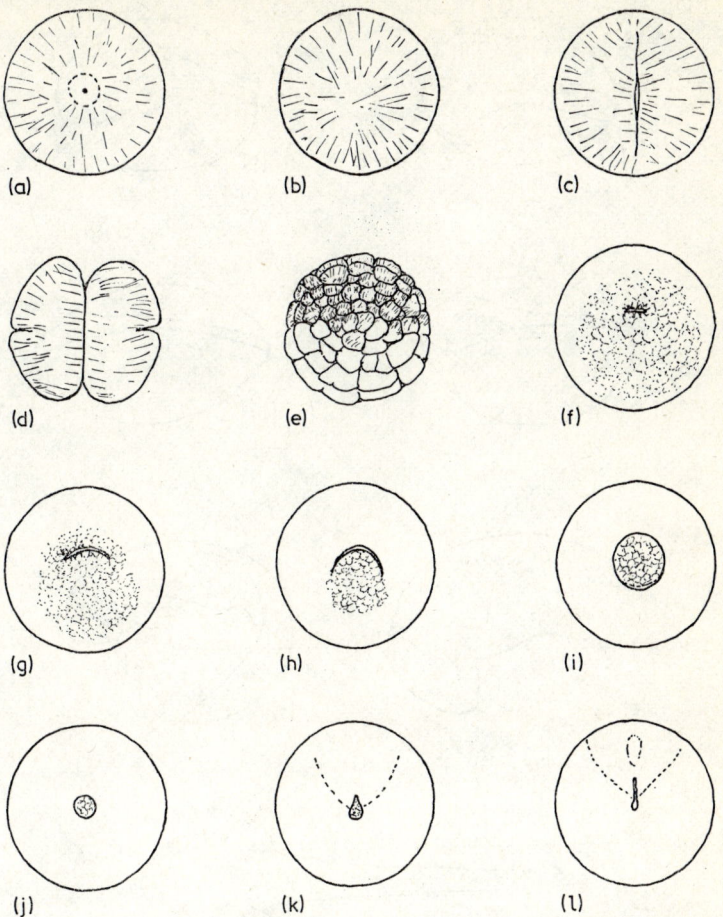

Fig. 6.5. *Stages in the development of Ambystoma mexicanum* I. (a) Recently fertilized egg. (b) About 2 hours after fertilization (H. St. 1). (c) Beginning of first cleavage (6 h). (d) Beginning of second cleavage (7 h). (e) Mid-cleavage, about 50 – 60 cells (H. St. 6, 12 h). (f) Beginning of gastrulation (H. St. 10, 40 h). (g – j) Gastrulation (H. St. $10\frac{1}{2}$ – 12). (k – l) End of gastrulation and beginning of neural plate (H. St. $12\frac{1}{2}$ – 13, 60 h). (a – d) viewed from the animal pole. (f – l) viewed from the vegetal pole. H. St. refers to Harrison stages for *A. maculatum* (see Hamburger, 1960). Times are for development at approximately 15°C.

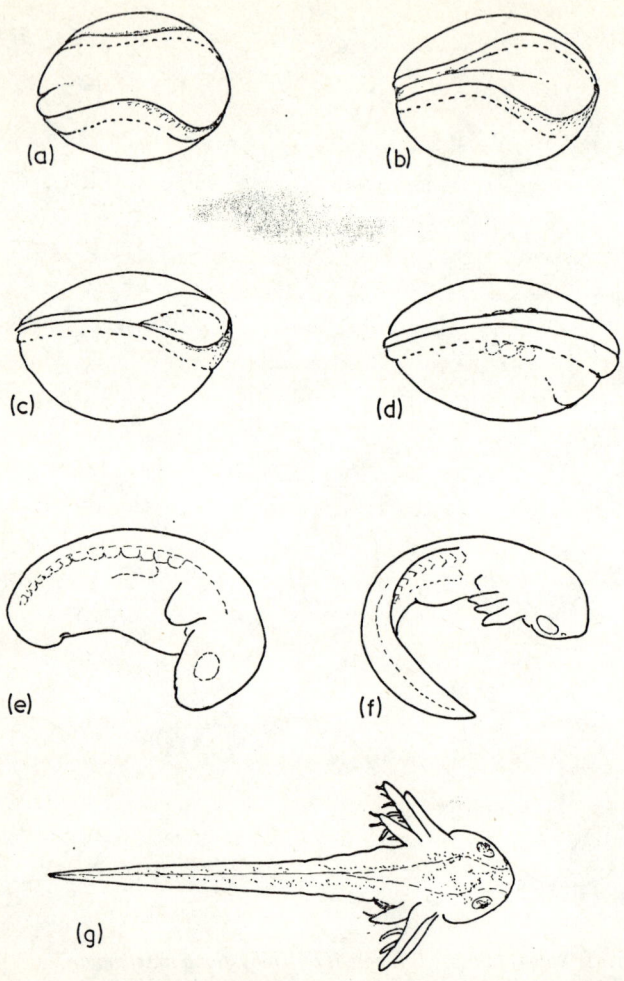

Fig. 6.6. *Stages in the development of Ambystoma mexicanum* II. (a – c) Formation and closure of the neural folds (H. St. 15 – 19, $3\frac{1}{2} - 4\frac{1}{2}$ days). (d) 3 somites (H. St. 22, 5 days). (e) About 15 somites (H. St. 30, 6 days). (f) Three gill rudiments and fore limb bud (H. St. 38 – 39, 12 – 13 days). (g) Newly hatched larva (c. 3 weeks). H. St. refers to Harrison's stages for *A. maculatum* (see Hamburger, 1960). Times are for development at approximately 15°C.

ments begin to appear. The following account draws attention to significant changes in the external appearance of the axolotl embryo as it develops into a recognizable urodele larva. A full stage by stage description is unnecessary as detailed Normal Tables, illustrating the development of urodeles, are readily available. (Hamburger, 1960; Rugh, 1962.)

Some 24 hours after the closure of the neural folds, that is at a stage corresponding to Harrison's stage 23 for *Ambystoma maculatum,* the embryo possesses a distinct head region in which the optic vesicles are apparent. In addition, about eight somites are present and just below the level of the fourth, the pronephric rudiment produces a small ridge. During the next 24 hours the features described above are emphasized and new ones appear. Thus, at about Harrison stage 28 there are about 15 somites, the line of the pronephric duct has lengthened and a pigmented otic pit has appeared just anterior to, and above, a distinct gill region. There is also an anal pit just in front of a newly formed tail bud (Fig. 6.6e). About three days after the closure of the neural folds the embryos are unmistakably larval in form (Harrison stage 35 to 36, Fig. 6.6f). At this stage the fore limb is just visible behind the gill region which now possesses three distinct rudiments. If these rudiments are examined carefully circulating blood will be seen and the heart itself can be detected pumping vigorously in the ventral region between the gill areas. Other signs of life are whole body twitches which are stimulated when the embryos are handled. Further development, particularly growth of the tail, elongation of the fore limb rudiments, and elaboration of the three gill rudiments and the head structures, occurs during the following week, towards the end of which, hatching occurs. A newly hatched larva is shown in Fig. 6.6g. Development does not, of course, stop with hatching but its further detailed description is beyond the scope of this book. However, attention must be drawn to the formation of the limbs which illustrates the tetrapod pattern so well. In urodeles the fore limbs develop before the hind limbs. This difference is so marked that the hind limbs are merely buds at the stage when their counterparts at the anterior end of the body are miniature, but fully functional limbs.

Effect of Temperature on the Development of Axolotl Embryos
Temperature has a very marked affect on the rate of development of axolotl embryos. This is shown in Fig. 6.7. For instance, at 20°C it takes just 24 hours to reach the beginning of gastrulation and at 7°C it takes seven days. This wide range is extremely useful when planning experiments.

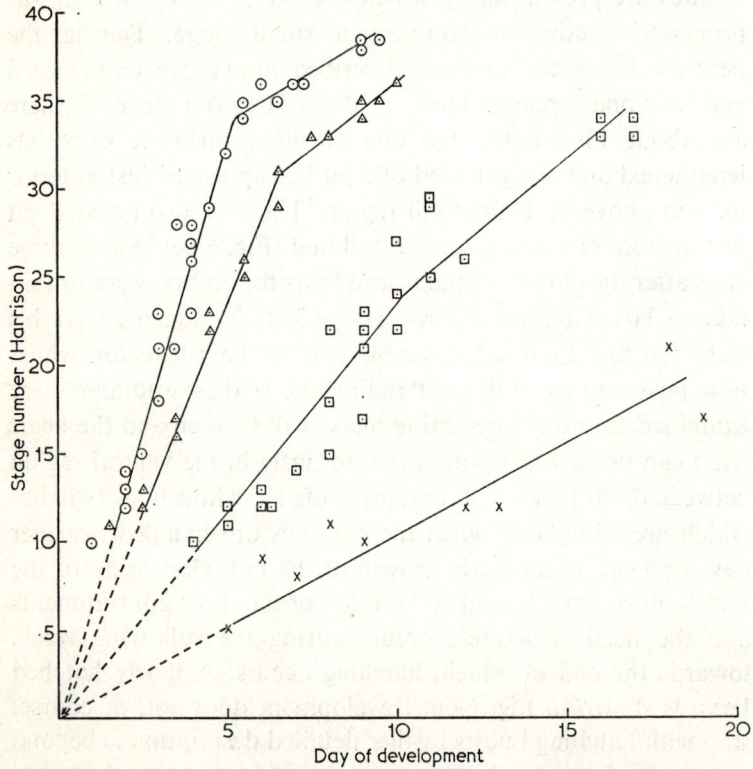

Fig. 6.7. Effect of temperature on the development of Axolotl embryos. −⊙−⊙− 20°C; −△−△− 15°C; −□−□− 10°C −×−×− 7°C, Stage numbers correspond to those of Harrison for *Ambystoma punctatum* (Billett and Adams, unpublished observations).

Notes on rearing axolotls from the larval stage

After hatching, the young larvae will survive up to a week without food, but as soon as the yolk has disappeared from the gut they will need feeding. Initially, for a period of about four weeks, they can be fed either on newly hatched *Artemia* (see p. 76) or microworms or a mixture of both. It is important to feed the young larvae adequately. If underfed they will become sluggish, very susceptible to fungal attack and will quickly die. At this stage the water will need to be changed fairly frequently, about every other day if *Artemia* is being used as the sole source of food. Attention should also be paid to the concentration of larvae: something of the order of 50 to 100 animals per 500 ml of water is maximal. After a month, when the larvae should be about 2 cm long they will take small *Tubifex*, and during the second month the diet should be changed gradually until it consists entirely of this worm. We do not recommend the use of chopped *Tubifex* for obvious reasons, but prefer to use the smallest worms when the axolotl larvae are small. These are obtained by shaking out a clump of *Tubifex* in a small volume of water. As the larvae grow they should be kept in a larger volume of water, and towards the end of the third month attempts should be made to feed the animals on small pieces of beef heart or liver. If the pieces of meat are left in the dish the animals will quickly learn to pick them up, and the tedious business of hand feeding at a later stage can be avoided. By the end of six months the animal should be on a diet consisting almost entirely of meat, but during the intervening period alternative feeds of *Tubifex* and meat should be given.

Experiments on Amphibian Embryos

Before describing individual experiments in detail, and to avoid repetition, a number of points which are generally applicable need to be emphasized. To begin with, it is essential to plan ahead by carefully reading through a particular practical, noting the requirements for animals and materials, and deciding on a rough timetable of events which will have to precede the actual operations. Such questions as 'how many embryos?' and 'what stages are required?' need to be asked, and disasters, such as finding sterile solutions and instruments too hot to handle because the embryos have developed more quickly than expected, need to be avoided. Unnecessary hold ups can also be avoided by paying attention to the quantity of glassware, instruments and solutions, which will be required. It is a good idea, having read the schedule, to make an estimate of what is required and then double it. Once the experiments are under way, sterile conditions in and around the immediate working area should be maintained as far as possible. In this connection, and also because the co-ordination of hand and eye seems to improve after a settling in period, it is an advantage to keep working for a fairly long time, i.e. about two hours, and to avoid enthusiastic forays between cups of coffee.

Many of the experiments described in the following pages necessarily involve the removal of the protective jelly capsules and vitelline membranes from the embryos. Once this has been done, normal development will only take place in a balanced salt solution that is osmotically compatible with the tissues. Successful operations also depend on a solution which encourages a quick healing of wounds. Holfreter's and Steinberg's solution are commonly used for these purposes. Their composition is compared on page 121. There appears to be little to choose between these solutions, although the buffer selected for Steinberg's medium makes it a little easier to prepare, and in some experiments on earlier embryos it appears to give marginally better results.

Once the embryos have been decapsulated they must be handled carefully; in particular they should not be allowed to come into contact with air/water interfaces, where the surface tension will destroy them. In this connection even greater care must be taken with small excised fragments of embryos. Care must also be taken not to allow the temperature to rise excessively, either during the operation, or during any culture period which follows it. Heat filters may be required on light sources used with dissecting microscopes. We recommend that during recovery from an operation, and subsequently, a temperature of 15 to 18°C be considered optimal for amphibian embryos.

After some of the operations described below many of the embryos will develop abnormally. It is important to distinguish those abnormalities which have been deliberately produced, from those which have arisen accidentally. Abnormal development will occur if the embryos are handled roughly, or if the solutions have not been made up correctly. To distinguish between planned and accidental abnormalities, and to test both the reliability of the solutions and the inherent quality of the embryos used, it is always essential to allow a small batch of decapsulated embryos to develop under the same conditions as the operated ones. Apart from this kind of control, it is important to carry out mock operations whenever possible to ensure a more reliable interpretation of the results, for example, to make simple wounds at the site of a graft.

Composition and Comparison of Holtfreter's and Steinberg's Solution

	Holtfreter (1931)	Steinberg (1957)
$NaCl$	3.5 g	3.4 g
KCl	0.05 g	0.05 g
$CaCl_2$	0.1 g	—
$Ca(NO_3)_2.4H_2O$	—	0.08 g
$MgSO_4.7H_2O$	—	0.205 g
$NaHCO_3$	0.2 g	—
1% N HCl	—	4.0 ml
Tris	—	0.56 g
Dist. H_2O	1000 ml	1000 ml
pH	7.5	7.4

Holtfreter's solution The $NaHCO_3$ is dry sterilized and added to the previously sterilized solution containing the remaining salts.

Steinberg's solution Although the recommended amounts of the tris/HCl buffer should give a pH of 7.4, some adjustment may be necessary before the solution is sterilized.

For original references, see Hamburger (1960).

To preserve embryos or larvae at the end of an experiment we recommend Smith's or Bouin's (aq.), fixatives (See Hale, 1958) and storage in 70 per cent ethanol. If sections are required the usual procedure leading to wax embedding should be followed. For most purposes sufficient histological detail is obtained from 5 to 7 μm sections stained with Ehrlich's haematoxylin and eosin. If a suitable microtome is available an alternative procedure, which gives much better cytological detail, is to fix in a formaldehyde/glutaraldehyde mixture, followed by embedding in araldite, or spur resin. Sections may then be cut at 1 μm and stained with toluidine blue (see Billett, 1968).

Unless specifically stated all the experiments described below refer to *Xenopus*, although they can all be adapted for, and were in fact first performed on, other amphibian embryos.

Artificial fertilization
For some investigations on the early development of *Xenopus* embryos it is necessary to know the exact time of fertilization. This is achieved by obtaining eggs from a female previously injected with chorionic gonadotrophin and fertilizing them with sperm from a crude homogenate of testes. The method described below broadly follows that of Gurdon (1967), although it differs from it in several details.

Mature females should be selected for this practical, and if they are in good condition only two or three will be needed to ensure a plentiful supply of eggs. The practical needs to be planned several days in advance, as the females are subjected to the injection procedure already described to induce ovulation (p. 99), that is, a priming dose of about 100 international units of gonadotrophin is given on the afternoon of the first day followed by a heavier dose, 200 units, on the second day. The females should be kept isolated once they have been injected. Care should be taken not to inject too much hormone as a generous dose may result in a wasted spontaneous ovulation. As described before, a reddening and swelling of the cloacal papilla indicates a response to the hormone. Eggs are obtained on the morning after the second injection by applying gentle pressure on the dorsal body wall, in such a way as to squeeze the eggs down the oviduct and out through the cloaca. The eggs are squeezed directly into an empty solid watch glass of about 2 to 5 ml capacity. The eggs must not be allowed to come into contact with water at this stage as this will cause their jelly coats to swell and will quickly prevent fertilization. About 20 to 30 eggs should be placed in each container. Fertilization is effected by adding about 0.2 ml of concentrated sperm to each group of eggs. The eggs are well mixed with the sperm preparation and after five minutes the mixture is transferred to a large petri dish containing 5 to 10 ml of water. If the fertilization has been successful the eggs will rotate freely and the first cleavage should be seen within two hours if normal laboratory temperatures are maintained.

The sperm preparation is obtained by killing a mature male and removing one of its testes. It is a useful practice to give the male an injection of chorionic gonadotrophin 24 hours beforehand to ensure an active sperm preparation. The animal is killed by a single injection of nembutal. Once the testis has been removed, it should be gently broken up by means of a small, all glass homogenizer containing about 0.5 ml of diluted Steinberg solution (1/10th normal strength, see p. 121), or chlorine free tap water. If the homogenizer is used too vigorously the tails will be broken off from the rest of the sperm and the preparation will be useless. Consequently, at this stage it is useful to examine a small drop of the preparation to see if intact and active spermatozoa are in fact present. We find it best to use the preparation as soon as possible, although it will remain active for at least an hour if kept in a cool place.

This preparation can be used for further studies, such as, the effect of diluting the sperm suspension and investigating ways of prolonging its useful life. Additional studies can also be undertaken using the eggs. How long for instance can they be left in the 'dry' state? Will a hypertonic salt solution prevent the jelly coat swelling and facilitate fertilization? Once the technique has been mastered it also becomes feasible to attempt more ambitious experiments. These include the production of gynogenetic haploids by treating the eggs with sperm which have been rendered genetically inactive, but not immobile, by u.v. or chemical treatment. Attempts can also be made to activate eggs parthenogentically by the time honoured device of pricking them with a fine glass needle.

Mechanical decapsulation
In almost all cases the protective jelly coats and vitelline membrane of amphibian embryos must be removed before any operation on the embryos themselves can be carried out. In this section we describe what is usually called mechanical decapsulation, i.e. the use of instruments, as opposed to chemicals. The preferred instruments for mechanical decapsulation are two pairs of watchmakers forceps; the task is made much easier if the points of the forceps are sharp and well matched. The operation must be carried out under a dissecting microscope at a magnification of about $\times 10$. The first stage is carried out in water and consists of pulling away as much as possible of the soft, rather sticky, outer jelly coat, preferably until only the tougher inner coat remains. To complete the decapsulation, the embryos are transferred to twice normal strength Steinberg or Holtfreter solution, in which the rest of the jelly coat and the vitelline membrane are removed. A good way to remove the inner coat is to penetrate and grasp it with one pair of forceps and then, using the other pair of forceps in a similar way, to complete the disruption of the coat by a fairly sharp pull. A similar operation can be used to remove the vitelline membrane. It is, however, difficult to lay down any exact guide lines for a procedure which is essentially a matter of practice and patience. Success will come as coordination of hand and eye improves. Embryos are likely to be damaged in the process and this is the reason for carrying out the last stage of the procedure in the solutions suggested, since they allow the embryos to heal rapidly. As with the

chemical procedure, it is a good idea to decapsulate the embryos an hour or two before they are required. This allows a good stock to be built up and time for the damaged embryos to heal. The above account applies particularly to *Xenopus* and the procedure may vary from one type of embryo to another. In the case of Axolotls, for instance, it is best to remove the outer jelly coat by simply rolling the eggs on a piece of filter paper (p. 112).

Chemical decapsulation
The mechanical decapsulation of amphibian embryos is a time consuming and somewhat difficult operation. Even after considerable experience it may take an hour to remove completely the surrounding membranes of about twenty early stage embryos. However, it is sometimes necessary to produce decapsulated embryos on a large scale, for instance, for biochemical analysis, or to provide an inexperienced class with large numbers of embryos for an experiment. In such cases chemical decapsulation can be carried out and we recommend the method of Spiegel (1951). The following procedure is based closely on Spiegel's method and has been found to work very successfully with *Xenopus laevis* embryos. Two conditions are essential for success: the papain must be freshly prepared and activated, and the pH of the reagents carefully adjusted to the stated values using a pH meter. The preparation of the reagents used in the procedure is described below.

Papain An appropriate amount of papain is weighed out and placed in N/5 Steinberg solution to give a 3 per cent w/v suspension. After shaking vigorously the mixture is left in a refrigerator overnight (*c*. 16 hours). The following day the suspension is filtered through a standard Whatman 40 paper (or an equivalent grade). If a large filter funnel, of about 15 cm diameter, is used and the filter paper is changed occasionally, it takes about one hour to filter 200 ml of the suspension. To activate the enzyme 30 mg of cysteine hydrochloride is added to each 100 ml of filtrate and when the compound has dissolved, the mixture is carefully adjusted to pH 6.7. The addition of the cysteine hydrochloride renders the solution acid, i.e. about pH 5.0, and several ml of 0.1 M NaOH per 100 ml of filtrate are usually required to bring the pH to the required value. Once activated the papain is effective for several hours, but it is best used immediately, and although it can be reactivated by adding more cysteine hydrochloride, it is better to make a fresh suspension when required.

Thioglycollate. A 2 per cent w/v solution is prepared by dissolving the appropriate amount of sodium thioglycollate in a given volume of N/5 Steinberg solution. The pH of the solution is adjusted to 8.5 using 0.1 M NaOH; about ten to twenty drops of the alkali is usually required. Although it is best to prepare the solution immediately before use, it may be stored in a refrigerator for a considerable period without loss of activity.

Practical details are as follows:
100 to 200 embryos are transferred to approximately 20 ml of the papain extract contained in a large 9 cm diam. petri dish. For this and subsequent transfers a wide mouthed pipette should be used. If the embryos are in a compact mass they should be broken down into smaller clumps to facilitate the attack of the enzyme. The embryos are left in the papain solution for one and a half to two hours. During this period the container should be rocked occasionally and the progress of the digestion of the jelly coats observed. With active enzyme the outer jelly coat quickly loses its stickiness and within half an hour the inner coat becomes opaque. Within an hour or so the outer jelly becomes quite soft and partially detached from the inner coat. Before treating all the embryos with thioglycollate the effect of the papain should be tested. A few embryos are removed from the papain, washed twice in N/5 Steinberg solution, and transferred to the thioglycollate solution. If the enzyme treatment has been successful the jelly coats will show signs of disintegration within a few minutes and after ten minutes they will be removed completely. The bulk of the embryos can then be treated in the same way, but if no disintegration of the jelly coat occurs then embryos should be

left in the papain for a further half an hour and the thioglycollate treatment repeated. At the end of the procedure it is essential to wash the decapsulated embryos several times with N/5 Steinberg solution.

Exogastrulation

If an amphibian blastula is placed in a hypertonic salt solution the blastocoele will collapse and the normal gastrulation movements will be prevented. Under these conditions the prospective chordamesoderm and the yolky endoderm evaginate away from a wrinkled sac of ectoderm (Fig. 6.8). In these exogastrulae, as they are called, the chordamesoderm does not make contact with the ectoderm and consequently no neural tissue is formed. Experiments of this kind were first performed by Holtfreter (1933) using urodeles, but it is possible to produce a similar effect with *Xenopus*. As the

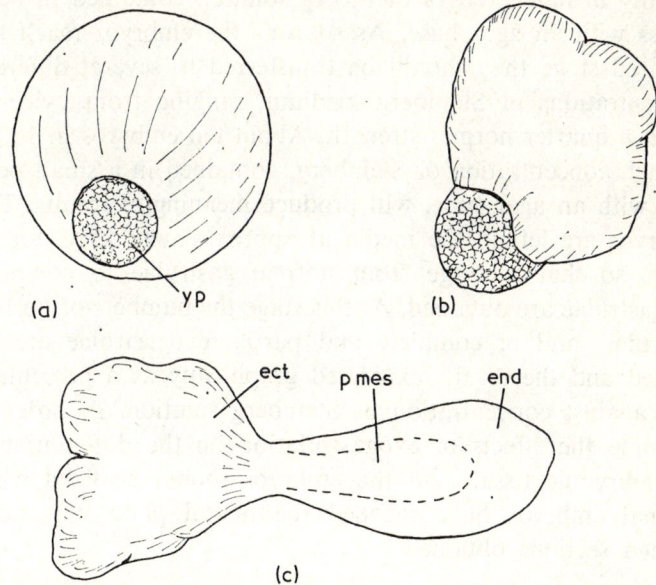

Fig. 6.8. *Exogastrulation in amphibian embryos* (*Xenopus laevis*) (a) Normal yolk plug stage. (b) Protruding yolk plug, partial exogastrulation. (c) Total exogastrulation. *Abbreviations*: end – endoderm; ect – ectoderm; p mes – presumptive mesoderm; yp – yolk plug.

production of exogastrulae does not involve much handling of the embryos the chances of success are good, and the experiment is therefore a useful one for an introductory course. It also draws attention to the need to control the molarity of the medium and emphasizes the importance of the contact between chordamesoderm and presumptive neural tissue during the early development of vertebrate embryos.

In view of their rapid development, it is necessary to treat *Xenopus* embryos with a hyptertonic salt solution at a relatively early stage in order to produce exogastrulae. One should aim to commence the treatment as soon as the embryos have reached an early blastula (preferably Nieuwkoop and Faber, stage 7). Thus to obtain material for the experiment the embryos must be decapsulated, either chemically or mechanically, at the earlier cleavage stages. Once free of their membranes the embryos should be well washed and placed initially in half strength Steinberg solution contained in petri dishes with an agar base. As soon as the embryos reach the required stage they should be transferred to several different concentrations of Steinberg medium, ranging from twice to about a quarter normal strength. About ten embryos in 10 ml of each concentration of Steinberg, contained in a small petri dish with an agar base, will produce meaningful results. The embryos are left in the media at approximately 15°C for 12 hours so that a range from normal gastrulae to complete exogastrulae are obtained. At this stage the numbers of normal gastrulae, and of complete and partial exogastrulae are recorded and the results expressed graphically as a percentage plot against concentration of Steinberg solution. In order to examine the effects of exogastrulation on the differentiation of embryonic tissues, all the embryos should be fixed when normal embryos have reached the neural plate stage, and stained sections obtained.

Ectodermal explants from gastrulae
During gastrulation in vertebrate embryos, the presumptive chordamesoderm makes contact with that part of the over-

lying ectoderm which is destined to become the neural plate. This contact allows the primary inductive interaction to take place and leads to the formation of an organized central nervous system. If this contact is prevented, as in exogastrulation (see above) then no differentiation of nervous tissue will take place. It is fairly easy to remove large pieces of ectoderm from amphibian gastrulae, and to culture them for a week or more. In this way the capacity of the isolated ectoderm to differentiate before or after an inductive stimulus can be demonstrated. Such explants have been widely used to study the nature of the inductive process (see for example, Yamada, 1960, 1962). The following experiment is designed to compare the differentiation of gastrula ectoderm which has, and has not, been subjected to an inductive stimulus. The preparation of explants is useful also in that it involves a partial dissection of an amphibian gastrula and thus enables the student to appreciate its structure.

The method of preparing the explants from *Xenopus* embryos is based on that described by Billett and Brahma (1960) which itself followed well established procedures. The aim is to make dorsal explants from the ectoderm covering the archenteron roof, and ventral ones from the prospective belly region of the embryo. The gastrulae are prepared for the operation by first removing their jelly coats and vitelline membranes. This can be done either chemically or mechanically but, if chemicals are used, then decapsulation must be performed at an early stage so that gastrulation does not occur during the procedure. Time must also be allowed for the removal of any damaged embryos.

To prepare the explants, several gastrulae are placed in a small petri dish (with an agar base) containing normal strength Steinberg medium, with 100 μg/ml of sulphadiazine added. Although any gastrula stage may be used, it is easiest and most instructive to start with mid gastrula stages. In these, the position of the blastocoele is often easy to detect as a large hump on the surface of the embryo, and the ectoderm comes away fairly cleanly from the roof of the forming archenteron. A small, shallow pit, sufficient only to hold the embryos in position, is made in the agar in the central region of the dish and the selected embryo manoeuvred into it. For the dorsal explant the embryo should be orientated so that its dorsal lip points away from the operator and the ectoderm covering the blastocoelic space should be towards him. To obtain the explant a transverse cut is made into the remains of the blastocoelic space using tungsten needles, followed by forward cuts on each side, to produce a broad flap hinged in the vicinity of the dorsal lip. The flap is raised gently from the region of the archenteron roof and removed from the embryo by making a second transverse cut, just above the dorsal lip. The location of the cuts is shown in Fig. 6.9. Once free, the excised portion is pushed away from the area of the operation and, if necessary, its ragged edges are trimmed with the needles. At this stage it is

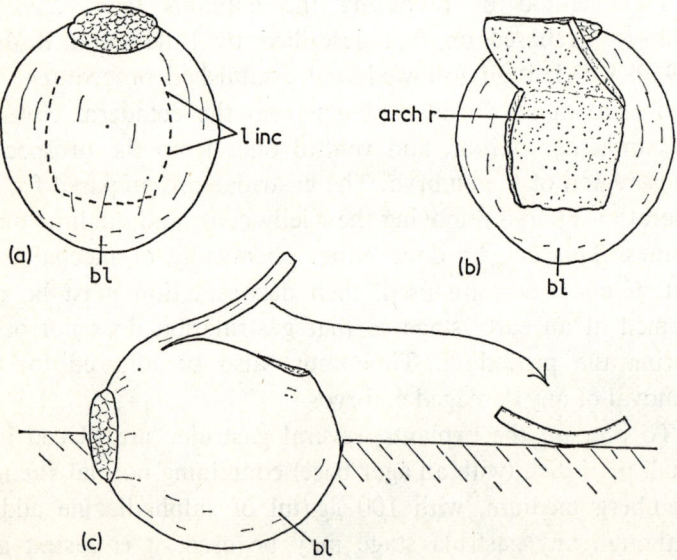

Fig. 6.9. *Preparation of dorsal ectodermal explant from an amphibian gastrula.* Scheme of operation viewed from above (a – b) and from the side (c) *Abbreviations*: arch r – roof of archenteron; bl – position of blastocoele; l inc – line of incision.

also necessary to remove, by very gentle use of a fine pipette, any unwanted material such as bits of yolky endoderm which may have been displaced and adhered to the inner surface of the explant during the operation. To make ventral explants the embryos are placed in a slightly different position, so that the ectoderm covering the blastocoelic space is central. In this case the longitudinal cuts are made in the reverse direction and the final transverse cut is made towards the back of the embryo. With practice one can 'skin' most of the embryo and make both kinds of explants from the same specimen.

As the explants are made they should be removed to another petri dish also containing normal Steinberg solution and with an agar base. Following the first operation a second embryo is manoeuvred into a fresh pit and another explant prepared, and so on.

After one to two hours in the 'recovery' dish the explants should be examined. Those which are likely to survive will, by this time, have rolled up with the original outer ectoderm surface still on the outside, and they should be transferred carefully to another petri dish containing normal strength Steinberg solution. Provided there are only a few explants in each dish there is no need to change the medium subsequently. It is best to allow the explants to develop at a fairly low temperature, preferably in the region of 15°C.

Under favourable conditions the explants normally survive for about a week, the ventral ones usually lasting even longer than this (Fig. 6.10). As time passes, the external appearance of the explants changes, indicating cellular differentiation in the isolated ectoderm. Details of these external changes, as well as the internal ones, are described fully elsewhere (Billett, 1968), but it is worth reiterating the main points here. In the case of the dorsal explants, the first obvious change is the appearance of the adhesive gland (see Nieuwkoop and Faber, stage 15) towards the end of the second day. Signs of ciliation appear during the third day, when some of the explants will begin to move on the surface of the agar. Changes indicative of neurulation occur during the fifth and sixth days

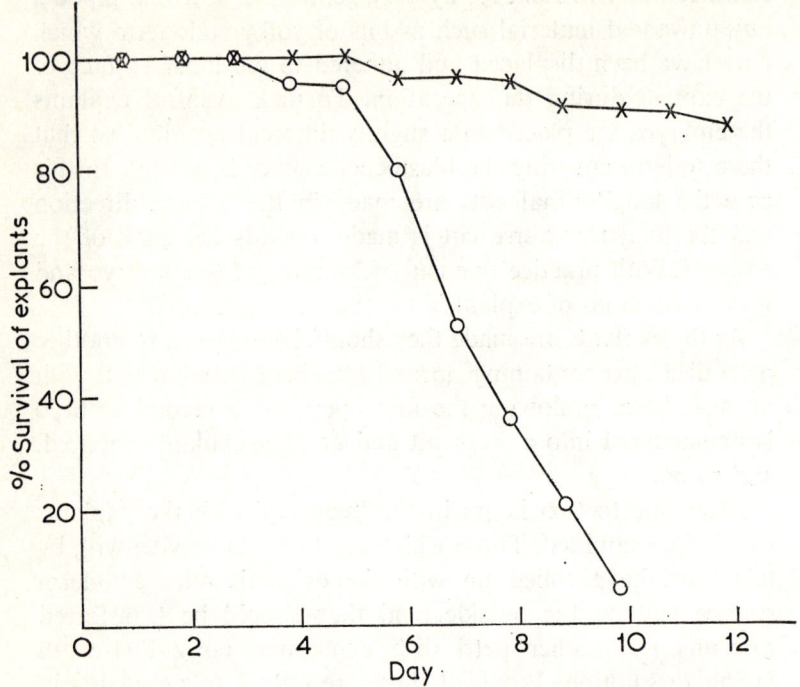

Fig. 6.10. *Survival of ectodermal explants from Xenopus laevis.* The explants were cultured at 15°C. –o–o– Dorsal (20 explants); –x–x– Ventral (28 explants). (Adapted from Billett and Brahma, 1960).

when, in some cases, a single vesicle may be formed in which can be seen small dense masses of neural tissue and often a pigmented eye structure. The ventral explants also develop cilia but do not undergo neurulation, the ectoderm in this case becoming ectodermal in character. Degenerative changes leading to the break up of the explants usually begin to occur during the sixth to seventh day, in the case of the dorsal, and during the tenth to twelfth day, in the case of the ventral explants. If it is desired to study the cytological changes in detail the explants should be fixed during the first six days of culture. For relevant methods see p. 122. An example of the kind of differentiation which can occur is shown in Plate 6.2.

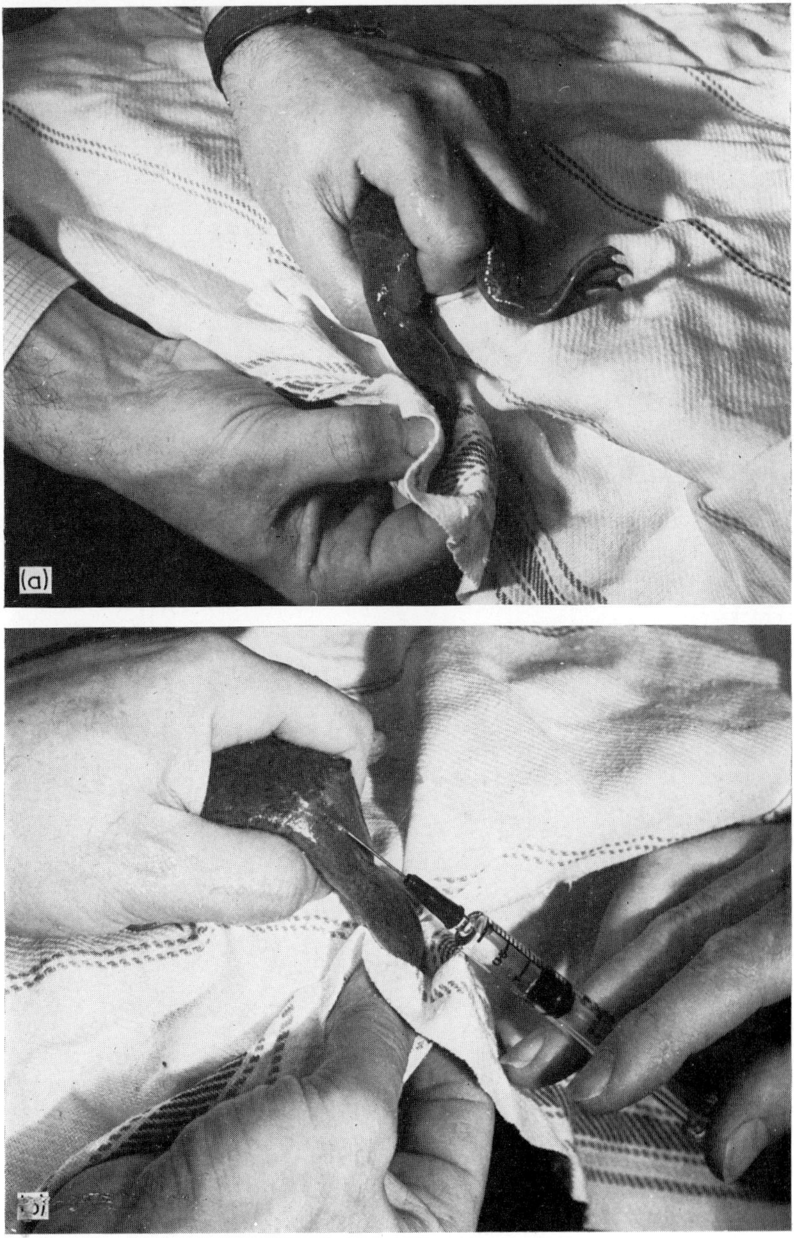

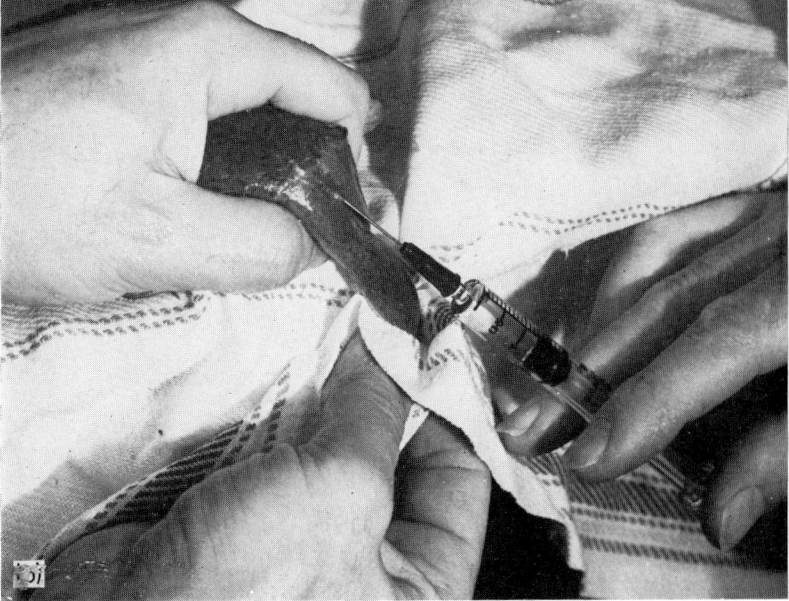

Plate 6.1 *Injection of Xenopus laevis*
(a) Method of holding the animal.   (b) The injection.

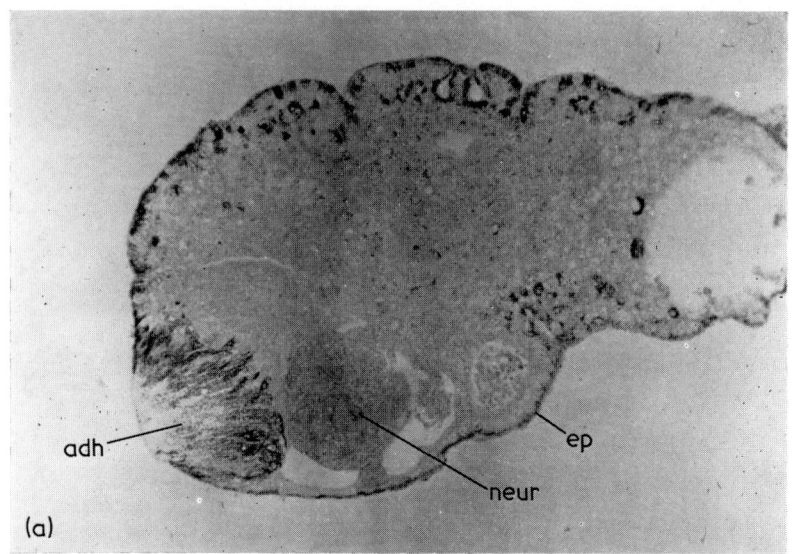
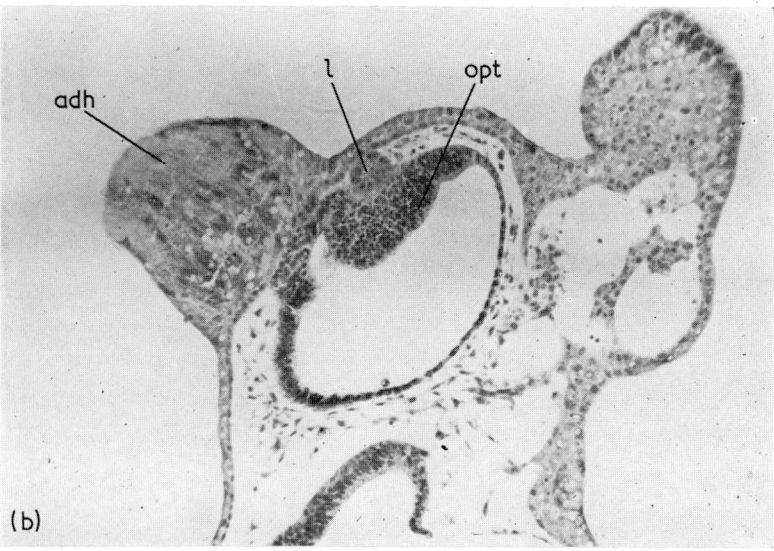

Plate 6.2 *Ectodermal explants of Xenopus laevis*
(a) Section of explant cultured for 60 hours. (b) Section of explant cultured for 72 hours. *Abbreviations:* adh – adhesive gland; ep – epithelium; l – lens-like structure; neur – area of neural differentiation; opt – optic vesicle (Billett and Brahma, 1960).

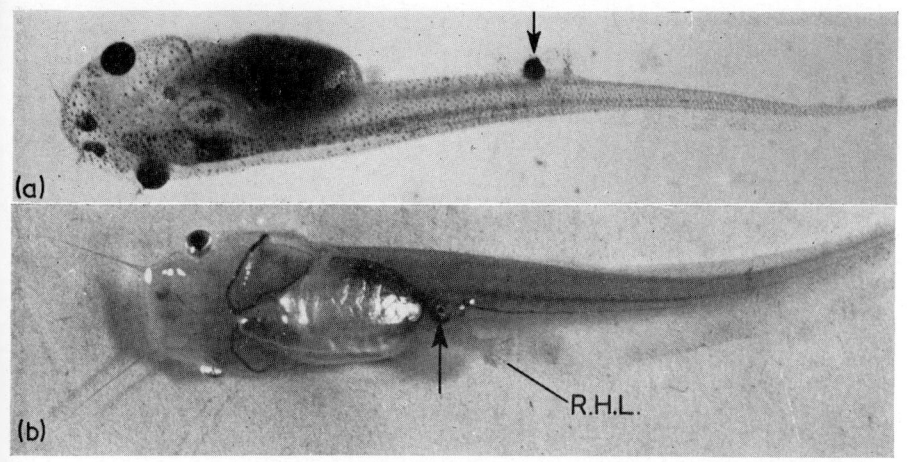
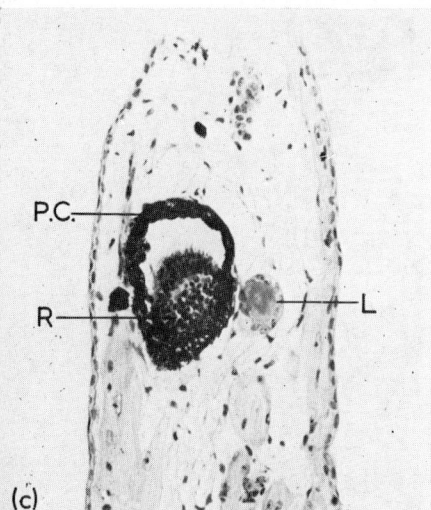
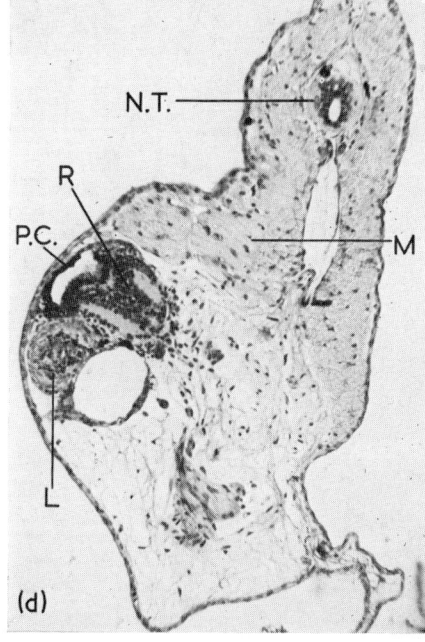

Plate 6.3 (a) Stage 50 *Xenopus* tadpole that had received an optic vesicle transplant when at the tail-bud stage. An eye (arrowed) has developed in the tail. (b) Stage 56 *Xenopus* tadpole that had received an optic vesicle transplant when at the tail-bud stage. An eye (arrowed) has developed in a site which should normally have been occupied by the left hind limb. The limb primordium was destroyed when making the transplant and only the right hind limb (RHL) has formed. (c) Longitudinal section through the tail of a *Xenopus* tadpole showing eye structures that have developed from an optic vesicle transplanted to the tail bud. The eye structures are smaller than normal and have an abnormal disposition. (PC – pigment coat; R – retina; L – lens) Stained with haematoxylin and eosin. (× 100) (d) Transverse section through the tail of a *Xenopus* tadpole showing eye structures that have developed from an optic vesicle transplanted to the tail bud. In this case the eye structures are better developed but still abnormal in position. (NT – neural tube; M – muscle; other abbreviations as in (c). Stained with haematoxylin and eosin. (× 100)

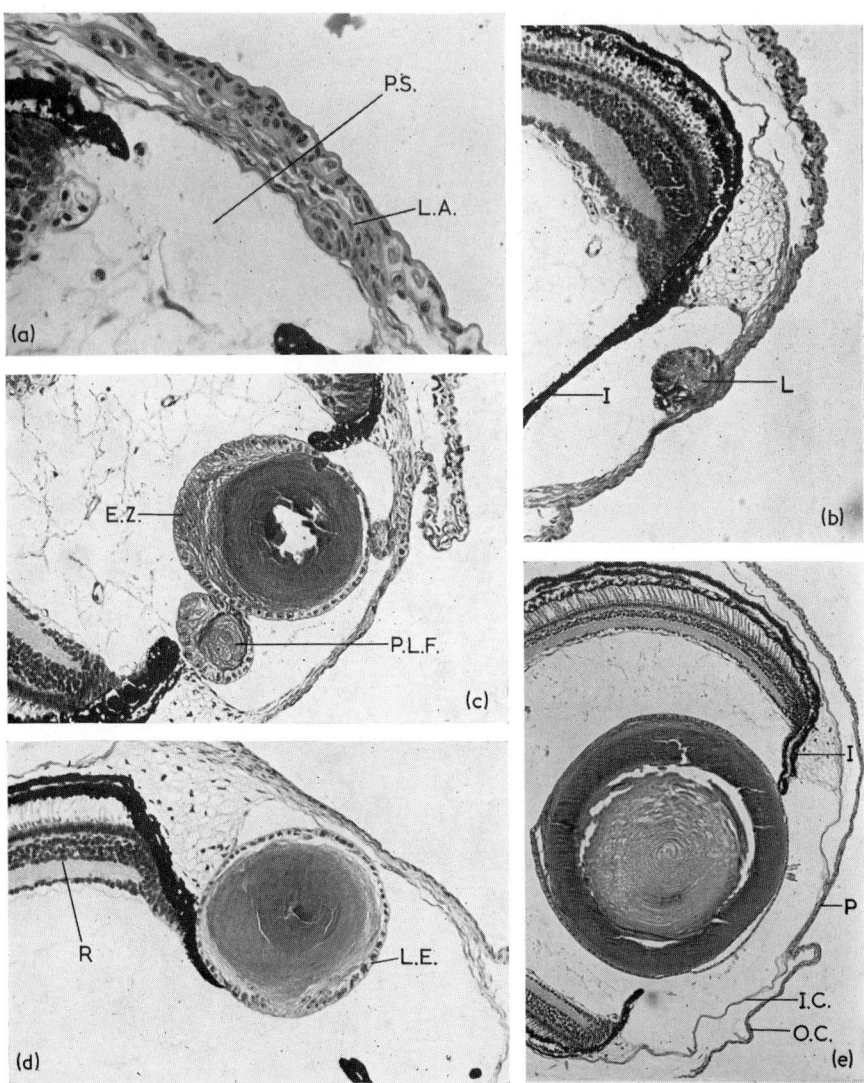

Plate 6.4 (a) Detail of a transverse section through the eye of a *Xenopus* tadpole 4 days after removal of the lens. A cell aggregate (LA) continuous with the inner cell layer of the outer cornea and typical of stage 2 of lens regeneration, can be seen overlying the pupillary space (PS). Stained with haematoxylin and eosin. (× 600). (b) Transverse section through the eye of a *Xenopus* tadpole 6 days after removal of the lens. An early stage 4 lens regenerate (L) can be seen still attached to the inner layer of the outer cornea and containing cells with prominent nucleoli. In this case the lens has developed more marginally, over the iris (I), rather than over the pupillary space. Stained with haematoxylin and eosin. (× 300). (c) Transverse section through the eye of a *Xenopus* tadpole 10 days after removal of the lens. This case is abnormal in that three lenses have regenerated – a small, middle stage 3 lens regenerate; a late stage 4 lens regenerate with characteristic primary lens fibres (PLF) formed into a nucleus; and a mid stage 5 lens regenerate in which the equatorial zone (EZ) is clearly visible and from which secondary lens fibres are forming. Stained with haematoxylin and eosin. (× 200). (d) Transverse section through the eye of a *Xenopus* tadpole 12 days after removal of the lens. A late stage 5 lens regenerate can be seen still attached to the cornea. (LE – lens epithelium; R – retina). Stained with haematoxylin and eosin. (× 200). (e) Transverse section through the eye of a *Xenopus* tadpole 16 days after removal of the lens and showing a fully formed lens lying between the margins of the iris (I). P – point of attachment of the inner (IC) and outer (OC) cornea. Stained with haematoxylin and eosin. (× 100).

Implantation of the dorsal lip of the blastopore

The discovery by Spemann that the amphibian dorsal lip had the ability to organize the basic structure of the embryo was of major importance in understanding the significance of gastrulation in vertebrate embryos. Spemann's original experiments involved removing the dorsal lip from one gastrula and grafting it into the prospective belly region of another. This resulted in the formation of a second embryo at the site of the graft. By using grafts between different species it was later shown that although the graft itself became incorporated into the induced structures, its key role was to cause the host tissues to organize themselves into the secondary embryo. Another method is to place the dorsal lip into the blastocoele of the host embryo before gastrulation starts. The implant then becomes trapped in the collapsing blastocoele and cell movements of gastrulation carry it round to the ventral region of the embryo. As before, a secondary embryo is induced in the vicinity of the implanted dorsal lip. This so called insertion technique has proved of immense value in the analysis of the primary inductive process (Saxén and Toivonen, 1962). Although it is feasible for students to graft the dorsal lip they will find it difficult to achieve unless they have had a great deal of experience in handling embryos. On the other hand, the implantation technique seems to create less difficulty, and we prefer to use it as the method for demonstrating the organizing ability of the dorsal lip.

As in the previous experiments these embryos should be decapsulated well in advance, and in this case it is essential to have a large number of the earlier stages in reserve, so that the embryos can be selected as they approach gastrulation. After decapsulation the embryos are placed in half normal strength Steinberg solution until they are required. The operations are carried out in normal strength solution contained in petri dishes with an agar base. Several embryos can be operated on in each dish.

Host and donor embryos are placed side by side in neighbouring pits in the agar layer and the donor is orientated with its dorsal lip facing the operator. The host is placed with the dorsal lip facing away, so that the roof of the blastocoele is clearly visible. A small area of tissue containing the dorsal lip is excised with tungsten needles and the isolated piece is left on the surface of the agar. Immediately after this a small transverse slit is made in the roof of the blastocoele of the host. The isolated dorsal lip is then taken up in a fine pipette and inserted into the opened blastocoele cavity. If the opening is not too big, the hole in the blastocoele roof will close fairly quickly and within an hour the embryos can be moved to another dish containing half normal strength Steinberg solution. The scheme of the operation is shown in Fig. 6.11. A successful implantation will produce a lump in the ventral region of the host embryo within 24 hours and a more organized structure will appear within 48 hours. Stained sections of the host embryos will reveal the extent of the experimentally induced structures. For control experiments pieces of ectoderm from the ventral region of the donor embryos should be implanted, and in addition, transplants of various parts of the blastopore lip from successive stages of gastrulation can be made. Experiments of this sort have demonstrated differences between head and tail organizer.

Optic vesicle transplantation in Xenopus laevis
In the normal course of the development of the vertebrate eye lateral outgrowths arise from the hind region of the forebrain. These outgrowths are called the optic vesicles and are destined to become the sensory retina and pigment coats of the eye; the connection made with the forebrain is destined to become the optic nerve. As the optic vesicles push out towards the overlying head epidermis they displace the intervening head mesenchymal tissue and start to flatten. On making contact with the epidermis, the optic vesicles invaginate and become transformed into double walled structures called the optic cups. These events are depicted diagrammatically in Fig. 6.12. It is the thicker invaginated wall of the optic cup which differ-

AMPHIBIA

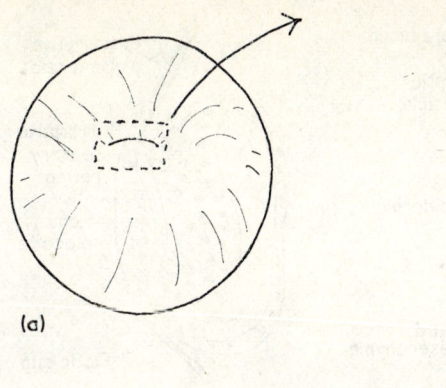
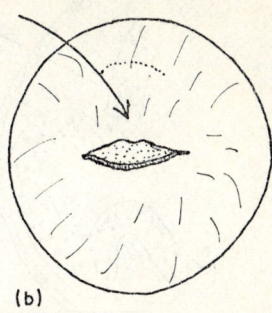
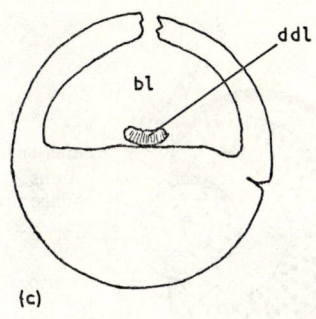
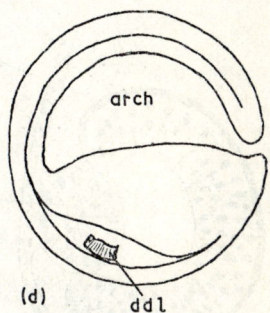

Fig. 6.11. *Implantation of dorsal lip of amphibian gastrula.* (a) Preparation of donor dorsal lip (ddl). (b) Preparation of host. (c – d) Diagrams showing the position of the implant at the beginning and at the end of gastrulation. *Abbreviations:* arch – archenteron; bl – blastocoele; ddl – implanted, dorsal lip of the donor.

entiates into the sensory retina, whilst the thinner outer wall develops into the pigment coat. The lens does not arise from the optic vesicle itself, but from the inner layer of the overlying head epidermis. An inductive influence which acts on the epidermis, emanates from the optic vesicle (and in *Xenopus laevis* also from the head mesoderm), causing a restricted part to become differentiated into a lens.

The purpose of the following exercise is to demonstrate that an optic vesicle removed from the head region and trans-

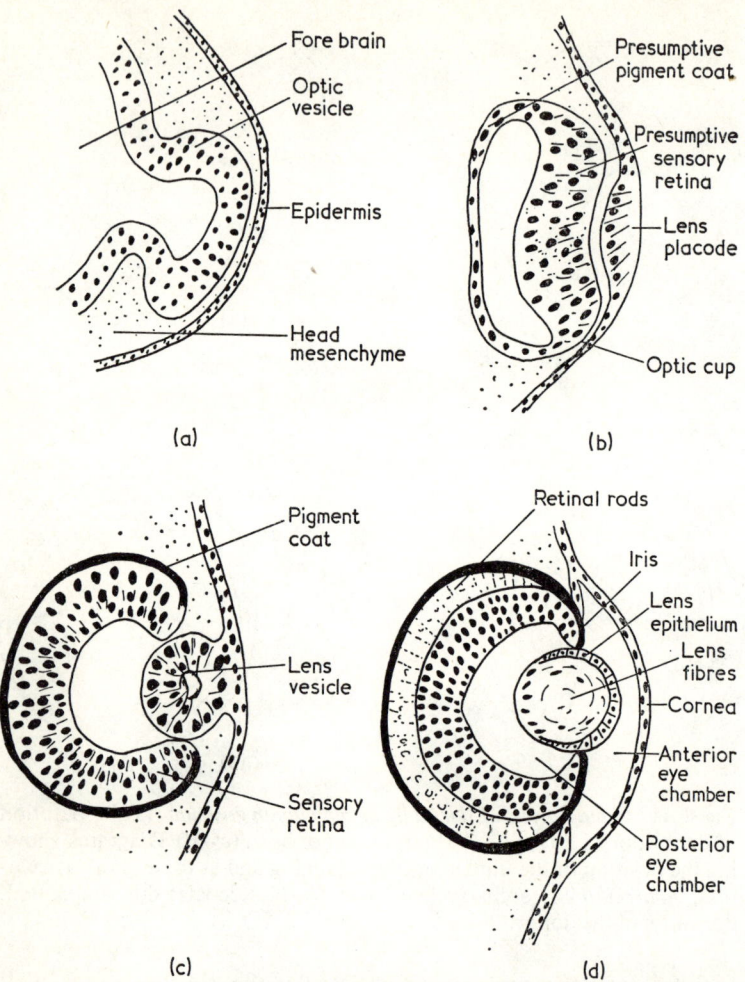

Fig. 6.12. Diagram showing successive stages in the development of the eye in amphibians. Outgrowth of the optic vesicle is depicted in (a), formation of the optic cup in (b), formation of the lens in (c) and the final arrangement of eye structures in (d).

planted to another site (in this case the tail region) still has the capacity to form normal eye structures, but forms no structures other than those of the eye. Furthermore, it can be demonstrated that epidermal tissue in the tail region, and in other regions of the body, will in many cases, form a lens if it comes under the inductive influence of the optic vesicle. This exercise requires a fair degree of manual dexterity and is suitable for advanced students who have time to gain expertise. The normal development of the eye should be studied first from relevant stained histological material. Practical details are as follows.

Xenopus laevis embryos at Nieuwkoop and Faber stages 24/25 and 28/29 (Fig. 6.2) are required. The earlier stage embryos are used as donors of the optic vesicles, which at this stage have not yet reached the overlying epidermis or exerted their inductive influence. The older embryos have formed a definite tail bud and are used as the recipients.

Animals are placed in sterile 10 per cent Holtfreter's solution and mechanically decapsulated (see p. 124) – a fairly easy procedure at this stage and one which does not damage the embryos. Together with a small volume of covering solution, donor embryos are transferred to a petri dish containing an agar base (1 per cent Ionagar No. 2 in 10 per cent Holtfreter's solution). A small cavity is cut into the agar in order to retain the embryo on its side during surgery and the petri dish is transferred to the stage of a dissecting microscope. An incision is now made by means of a tungsten needle in the pigmented epithelium overlying the optic vesicle and the epithelium peeled away to reveal the more lightly coloured optic vesicle dome. Two hair loops are now placed over the optic vesicle as depicted in Fig. 6.13. By pulling one hair loop away from the other, the optic vesicle is cleanly excised. An alternative procedure is to remove the optic vesicle with its covering of epidermal tissue intact and then to dissect away the epidermis. The transplant is now transferred by means of a fine pipette to a 1:200 000 solution of Nile blue sulphate in 10 per cent Holtfreter's. Nile blue sulphate is a vital dye; it is accumulated by cells, making them more easily visible, but it does not impair their normal function. About 15 minutes is required to stain the transplant adequately and during staining the recipients should be prepared. The recipients have

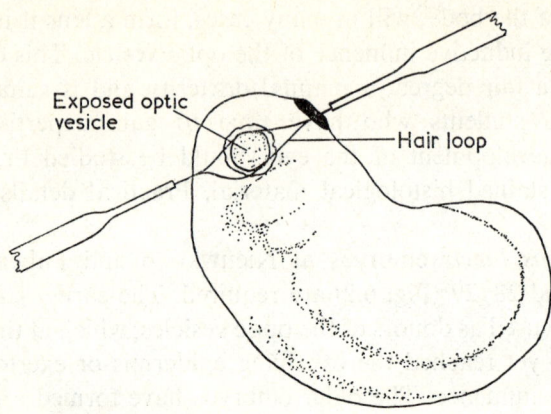

Fig. 6.13. Method of removing an optic vesicle from a stage 24 *Xenopus* tadpole by means of hair loops.

reached a stage at which the myotomes are capable of functioning and thus may display intermittent writhing movements, making implantation of the transplant difficult. In this event embryos should be placed in MS222 (1:2000 in 10 per cent Holtfreter's solution) to effect anaesthesia. They are then transferred in Holtfreter's solution to a fresh petri dish containing agar and placed in cavities so that they lie on their sides. The transplant should now be transferred so that it lies alongside the recipient. An incision is made in the tail bud with a tungsten needle and an oblique hole, large enough to accommodate the transplant, excavated. At this stage it will come as a surprise to most students, to observe how quickly cell movement, associated with wound healing, takes place. The transplant is manoeuvred into position with watchmaker's forceps and pushed into the transplant bed. The graft should be orientated in the same position it normally occupies during development, i.e. dome outwards. A small piece of broken coverslip can now be laid over the wound to hold the graft in position until the edges of the wound have grown over. Grafts are frequently rejected because repair takes place beneath the transplant rather than over it. It is thus essential to get the graft in position as quickly as possible. When the transplant has healed in, the embryo is transferred to 20 per cent Holtfreter's solution containing sulphadiazine (100 μg/ml), which is a bacteriocidal agent.

Embryos should be observed daily and the appearance of the transplant recorded. Within 48 hours evidence of eye structures are usually apparent when the transplant is viewed externally under a dissecting microscope, and after four or five days a refractile lens can often be observed. Plate 6.3a shows the appearance of an optic vesicle transplant in a tadpole tail at 26 days after transplantation. As in the case illustrated, the eye is frequently enclosed in a vesicle and is somewhat smaller than the normal eye since not all of the optic vesicle is removed by surgery. An interesting extension of this exercise is to subdivide the transplant into smaller pieces and to determine whether or not normal eyes are produced after transplantation. Students are often amazed to find that even quite small transplants have the ability to regulate and produce normal, although very much smaller, eyes. The eye is not functional since there is no nervous connection to the brain. Some of the successful transplants should be fixed in Bouin's, embedded, and sectioned, and detailed drawings made of the eye structures. Plate 6.3c and d shows the appearance of sections through eye vesicle transplants in tadpole tails.

It will be appreciated that the tail makes the best site for transplantation since there is little likelihood of interfering with the normal development and functioning of the tadpole. Sometimes transplants will end up in sites other than the tail through faulty technique. Plate 6.3b shows the result of transplanting an optic vesicle to what had been the site of the limb bud. A further advantage of making the transplant in the tail will be appreciated if the tadpole can be kept alive until metamorphosis. At metamorphosis the tail shrinks due to the high concentration of thyroxine circulating in the blood and acting on the tail tissues. Apart from causing the cornea to become thickened, thyroxine has no regressive effect on the eye contained in the tail. It simply gets carried back into the sacral region as the tail regresses. Striking pictures of this are shown in the paper by Schwind (1933) in work carried out on the frog (*Rana sylvatica*), and it demonstrates very strikingly the way the various tissues are primed to respond in different ways to the action of this hormone.

Other experiments on amphibian embryos
The experiments on amphibian embryos which we have described in detail reflect a personal choice based on their success with undergraduate classes. There are, of course, many other experiments which can form the basis of advanced undergraduate work or of individual projects. A short account of several such experiments is given below, together with an assessment of their feasibility. For further details see the reference list at the end of Chapter 1.

Vital Staining By marking small areas of the surface of late blastulae or early gastrulae with vital dyes the subsequent movement of these areas during gastrulation, and their fate at the end of the process, can be determined. It was in this way that the now well known presumptive areas were mapped out on the surface of the amphibian blastula (Vogt, 1929). Bebbington and Thompson (1967) describe the application of this procedure to *Xenopus* early gastrulae, and give details of the preparation of the dye markers (small pieces of agar soaked in methylene blue or neutral red). The process of neuralation and the fate of other areas, for example those from the neural plate or neural crest, can also be followed by using this method. The technique is more difficult than it sounds, and since it also requires much patient observation to achieve a satisfying result, we suggest that this kind of work is suitable only for small groups of advanced students.

Dorsal Lip transplants Spemann and Mangold's classic experiment (1928, see Spemann, 1938) of transplanting the dorsal lip of a gastrula into another gastrula of the same age, demonstrates the ability of this tissue to organize a secondary embryonic axis. This is an experiment we have tried several times as a class practical but without much success. It is certainly difficult to carry out using *Xenopus* embryos at room temperature, probably because they develop too quickly. Urodele embryos are larger and develop more slowly and so offer a greater chance of success. One of the best accounts of experiments which can be carried out in this field is given by Spemann (1938).

Removal of Pieces of Neural Crest. It is a relatively simple matter to remove small pieces of the neural crest from the neural fold stages of amphibian embryos and to see the effects of this operation. Horstadius and others performed many experiments of this kind and other related operations involving removal and re-orientation of whole areas of the neural plate and crest material. Such work showed that the neural crest was responsible for the formation of many important structures during subsequent development. These included cartilaginous components of the chondrocranium from the head crest, and melanocytes and elements of the spinal ganglia from the trunk crest. A comprehensive account of these experiments and their significance is given by Horstadius (1950). Since the effects of removing part of the trunk crest are easy to see, that is as a defect in pigmentation, this is a feasible practical exercise for an undergraduate class.

Nuclear Transplantation. Although of great importance in research, nuclear transplantation is obviously a technique which, in view of the preparation needed and the skill required, cannot be contemplated as a practical study, even for an advanced group of students. It is possible as an individual project provided sufficient time and adequate material are available. Details of the technique are given by King (1967).

Experiments on Amphibian Larvae

Although it is often difficult to produce embryos at the right stages of development for particular experiments, the same cannot be said about larvae. After every experiment requiring embryos there is usually a surplus of material which can easily be reared to larval stages. Several interesting and informative experiments can be performed on these animals, provided the Home Office requirements are borne in mind. These experiments fall into two groups; those involving regeneration and those involving metamorphosis. The two practical studies described below deal with lens regeneration and the effects of thyroxine on isolated larval tails, but before describing these in detail a brief indication of other possibilities is given.

The ability of the appendages and tail of amphibian larvae to regenerate is well known, and has been especially well investigated in Urodeles. Axolotls are excellent material for this kind of work. The simplest kind of experiment is to anaesthetize the larvae with MS222 and to snip off a portion of tail or part of a limb and then to observe the various stages and rate of regeneration. There are obvious variants of this procedure, such as altering the level and angle of the cut.

Xenopus and Axolotls are both suitable for experiments on metamorphosis. For instance, *Xenopus* larvae can be induced to metamorphose prematurely by treating them with small amounts of thyroid extract (1 part in 10^8 parts of water). Conversely metamorphosis can be retarded and completely stopped at pro-metamorphosis (stage 57) by treatment of younger larvae with a 0.1 per cent aqueous solution of propyl thiouracil (Turner, 1973). With respect to Axolotls, which normally remain and breed in a larval state, (a condition known as neoteny) metamorphosis can be induced by treating them with thyroxine or tri-iodothyronine. Working in our laboratory, Ingram (1969) was able to induce metamorphosis in one year old Axolotls, by giving them fortnightly intraperitoneal injections of 0.1 ml of a solution containing 100 μg/ml of tri-iodothyronine in Steinberg's solution. The treated animals metamorphosed completely about 10 weeks after the first injection.

Lens regeneration from the cornea in Xenopus larvae

The dorsal margin of the iris of larval and adult urodeles, belonging to the genus *Triturus*, has long been known to have the capacity to regenerate a new lens should the existing one be removed, although in the normal course of development the lens arises from the ectoderm as a result of the inductive influence of the closely apposed optic vesicle. This remarkable display of cellular metaplasia, in which certain cells of the iris lose their tissue specificity and acquire a new one, is usually referred to as 'Wolffian regeneration', after its discoverer Wolff (1895). He showed its occurrence in the european newt *Triturus*, although Colucci (1885) had apparently observed such regeneration before him (for references see review by Yamada, 1967). A similar phenomenon has more recently been shown to occur in larval *Xenopus laevis*, but in this case the lens more often than not regenerates from the cornea. The cellular events involved have been described in detail by Freeman (1963) to whom reference should be made for a discussion of the possible reasons for its occurrence in this species.

Students can soon acquire the skill needed to remove the lens without causing excessive damage and in our experience a high proportion of the tadpoles regenerate lenses. Sufficient time must be allowed for the subsequent histological procedures to be carried out. It must also be appreciated that the free living tadpole comes within the scope of the Cruelty to Animals Act, so that the necessary licence must be obtained before lens removal with subsequent recovery of the tadpole, can be performed. For these reasons this exercise is better suited for a small group of advanced students and it should be started at the beginning of a course in order that all the necessary histological work can be completed by the end of the course. The method is based essentially on that described by Freeman (1963), with minor modifications.

Animals

Xenopus laevis tadpoles obtained by the method previously described (p. 98) are staged according to Nieuwkoop and Faber (1967). In his original study Freeman (1963) used stages 46, 50, 56 and 66, representing the end of the embryonic period, the beginning and end of the larval period, and the juvenile stage. Because of its small size Stage 46 is not easy to work on and it is better to use stages 50 to 58 first. We find it best to divide the students into small groups, each working on a different stage.

Operative technique

Animals are anaesthetized with a 1 : 2000 solution of MS222 (Sandoz Ltd.) made up in 10 per cent Holtfreter's solution and then transferred to a petri dish which has had a layer of agar (1 per cent Ionagar No. 2 in 10 per cent Holtfreter's solution) added to form a base. A thin slice of agar is removed so as to accommodate the tadpole lying on its side; it is kept in position with a piece of microscope slide covering its lower half. All operations are carried out under a dissecting microscope. The orbit is carefully grasped with fine blunt ended forceps and an incision is made in the cornea with a sharp tungsten needle. It may be found easier to make this incision by passing a tungsten needle beneath the cornea, across the pupil, and out again and using a sharp scalpel blade (No. 11 or smaller) to cut the cornea against the needle across the pupil (Fig. 6.14a). Pressure is applied with forceps to the back of the eye, which forces the lens through the corneal incision. With the aid of forceps the lens is freed from its attachments and removed (Fig. 6.14b). The lens should be checked under the microscope to see that it is intact and that no part has been left behind. If this is not the case, or if haemorrhage and excessive damage occurs during the operation, then the tadpole should be discarded. If analysis of lens proteins is to be carried out at some subsequent stage (see p. 149) then lenses from the same stage tadpoles should be pooled in a small tube containing a drop of normal saline. The cornea of the other eye should be incised in the same way, to act as a control. The animals are reared individually in 250 ml beakers containing 10 per cent Holtfreter's solution and fed with Complan. After one to two days they can be transferred to aged tap water.

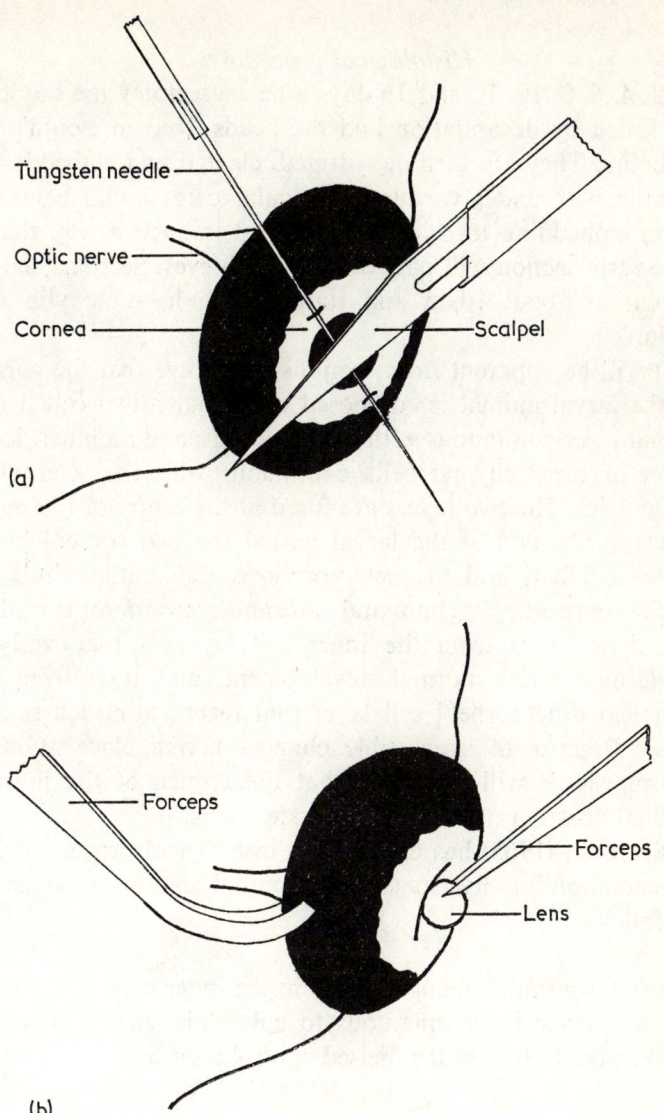

Fig. 6.14. *Removal of the lens from the eye of a Xenopus tadpole*. In (a) is depicted the method of making an incision in the cornea by cutting with a scalpel blade against a tungsten needle and in (b) is shown the method of extracting the lens.

Histological procedures

At 2, 4, 6, 8, 10, 12 and 16 days after lensectomy the tadpoles are killed by decapitation and the heads fixed in Bouin's for 24 hours. They are then dehydrated, cleared and embedded in paraffin wax under vacuum. Particular care should be taken when embedding to orientate the head in such a way that a transverse section will pass through both eyes. Sections should be cut at about $10\mu m$ and stained with haemotoxylin and eosin.

It will be apparent from sections of the eye that the cornea of the larval animal is composed of an outer two celled epithelial layer continuous with the epidermis, and an inner, loose sheet of mesenchymal cells continuous with the sclera (see Plate 6.4e). The two layers are fused at the centre of the pupil. Towards the end of the larval period the two corneal layers grow together, and at metamorphosis differentiate into the thicker corneal epithelium and *substantia propria* of the adult. The lens forms from the inner cell layer of the overlying epidermis during normal development, and it is from the corresponding corneal cell layer that regenerating lenses also arise. Because of irreversible changes taking place at metamorphosis, it will be found that the cornea of the juvenile animal does not normally regenerate a lens.

Freeman (1963) has divided the histological process of lens regeneration into five stages, the essential features of which are as follows:

Stage 1 An initial change occurs in the inner cells of the outer corneal layer from squamous to cuboidal. Such a change is also characteristic of the incised control cornea.

Stage 2 Two days following lensectomy cells of the inner layer of the outer cornea aggregate into a loose clump over the pupillary space (Plate 6.4a).

Stage 3 Four to five days following lensectomy the cells of the aggregate become orientated with respect to each other and form an epithelial vesicle (lentoid) which grows in size.

Stage 4 About six days after lensectomy cells of that part of the lentoid farthest from the cornea show a distinct enlargement of nuclei and nucleoli (Plate 6.4b). Such changes are the first morphological signs of specific lens protein synthesis. Mitosis in the lentoid becomes restricted to the periphery. The cells with large nuclei start to produce primary lens fibres and these are added to by other cells. Lentoids may be found attached or detached from the cornea at this stage.

Stage 5 About ten days after lensectomy the nuclei of the primary lens fibres start to disappear and secondary fibres grow from the equatorial zone of the lens epithelium (Plate 6.4c). The lens now increases in size but no major change in histology takes place. Some late stage 5 regenerates may still have the lens attached to the cornea (Plate 6.4d). Ultimately the regenerated lens occupies a size and position identical to that in the control eye (Plate 6.4e).

Each stage can be divided into early, middle and late, and regenerates are staged accordingly. If the lens regeneration stage is plotted against time after lensectomy, for each of the tadpole developmental stages used it will be observed that older tadpoles take longer to regenerate lenses compared with younger tadpoles (Fig. 6.15). This is due to the fact that it takes longer for the aggregate to form in older tadpoles and that a certain minimum lentoid size relative to the eye cup volume, has to be reached before lens fibre formation will occur (Freeman, 1963).

An interesting abnormality was discovered during our investigations and is illustrated in Plate 6.4c. Instead of a single regenerate, three have formed, all of which are at different stages of differentiation. It is impossible to tell whether all the regenerates formed at the same time and then developed at different rates, or whether they arose at different times after lensectomy and proceeded to differentiate at the same rate.

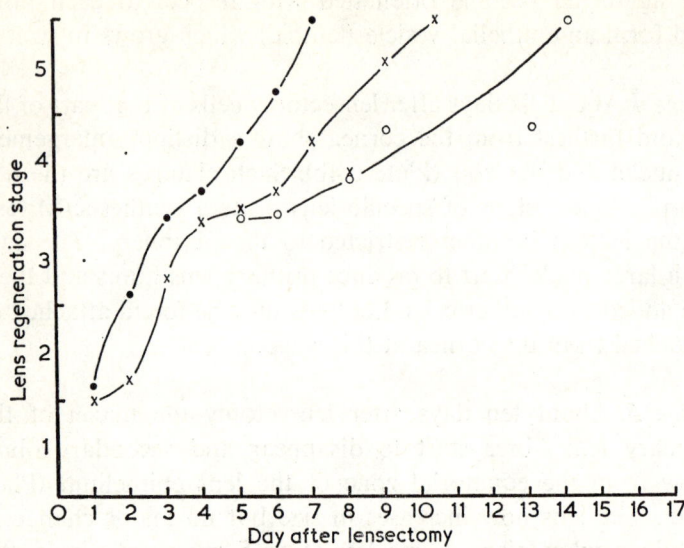

Fig. 6.15. Comparison of the rate of lens regeneration for *Xenopus* developmental stages 46, 50 and 56. (●——●) = stage 46; (×——×) = stage 50; (○——○) = stage 56. (From Freeman, 1963).

Whatever the case, this is a finding of considerable interest and it focusses attention on the mechanism of control of lens formation by the cornea. It raises the question: what prevents the cornea from differentiating further lenses when an existing lens is present? A worthwhile extension of this investigation is to remove the lens incompletely or to replace small fragments of it. In such situations it has been shown that the cornea does not normally produce a lens, indicating that the existing lens is exerting some form of control. If the cornea is removed and cultured in relatively simple conditions *in vitro*, lens formation can still take place (Campbell and Jones, 1968). However, if a lens is placed in close proximity to the isolated cornea, no lens differentiation takes place (Campbell, personal communication). Some factor(s) emanating from the lens presumably suppress lens formation by the cornea. In the abnormality illustrated, one can only suppose that the factor(s) have failed to be produced by at least two of the lentoids. Waggoner (1973) has recently shown that cornea transplanted to a lensectomized anterior eye chamber, or to the blastema formed from a regenerating hind limb, will form a lentoid, but that cornea transplanted to the dorsal tail fin will not.

Immunoelectrophoretic analysis of lens proteins during development of Xenopus laevis-a study of chemical differentiation

The adult vertebrate lens is rich in soluble proteins which are referred to as crystallins. Generally three classes of soluble crystallins (α, β and γ) have been recognized on the basis of their differing mobility on electrophoretic analysis. Thus at pH 8.6 the γ-crystallins migrate furthest towards the cathode, β-crystallins are less cathiodic, and α-crystallins have a slightly anodal mobility (Fig. 6.19). Improved methods of separating proteins have shown that all three classes of crystallins are heterogeneous and considerable attention has been paid by research workers investigating the mechanism of lens differentiation to the sequential development of such lens proteins (see review by Clayton, 1970). It will be appreciated that differentiation has to be considered not only in terms of changing gross morphology, but also in terms of the chemical changes underlying such morphology. The development of the lens from relatively unspecialized epidermal cells, to a structure composed of highly specialized lens fibres with unique proteins, provides us with a striking example of differential gene activity.

The purpose of this exercise, which is suitable for a small group of advanced students, is to analyse lens proteins at different stages of development of the tadpole, and to compare them with those of the adult lens.

The method of analysis makes use of a technique called 'immunoelectrophoresis'. This technique has led to great advances in the analysis of complex mixtures of proteins and is a relatively easy procedure to carry out. (For a general account of the method see Grabar and Burtin, 1964). The principle is as follows: an antiserum is first prepared by injecting the protein antigens present in the mixture to be analysed into a rabbit. Since the proteins (if they are *Xenopus* lens crystallins, for example) are foreign to it, the rabbit will produce specific antibodies (precipitins) to each of the proteins in the mixture. A solution of the proteins to be analyzed is subjected to electrophoresis in an inert transparent medium (Oxoid Ionagar No. 2, or other suitable material such as 'Agarose'). When electrophoresis is completed, the rabbit antiserum is placed in a lateral well (see Fig. 6.19) and allowed to diffuse through the agar. Where the separated protein antigens meet their specific antibodies in optimal proportion, a white precipitin line in the shape of an arc is formed. Such precipitin lines, indicative of single proteins, can be photographed, or stained after suitable treatment, and their position and number recorded. Practical details are as follows.

Preparation of adult lens crystallins

Adult *Xenopus* are killed by immersing them in 0.2 per cent MS222 in tap water, contained in a covered vessel. Death is ensured by decapitation when anaesthesia is complete. In order to extract the lens, the skin covering the orbit is grasped with forceps and downward pressure applied so as to cause the eyeball to bulge out. An incision is then made in the cornea with a scalpel blade. Pressure applied on either side of the eyeball with forceps will now force out the lens. Any attached tissue is carefully removed and the lenses are washed in 0.85 per cent saline. They are then transferred to a semi-micro glass homogenizer, and homogenized in a small volume (0.5 ml) of saline. The contents of the tube are then centrifuged in order to remove the insoluble protein fraction. During preparation of the antiserum about 5 mg of soluble protein is needed for each injection, this being the product of approximately eight adult lenses. Rather than kill adult *Xenopus* specifically for the purpose of obtaining lenses, it is clearly advantageous if a source of dead animals can be found. For instance in our department adult *Xenopus* have been required for their blood, and have also supplied the lenses. The total protein content of the lens crystallin solution can be conveniently determined colourimetrically by the method of Lowry, Rosebrough, Farr and Randall (1951), as described below.

Total protein determination

The reagents required are as follows:
(A) 2 per cent Na_2CO_3 in 0.1 N NaOH
(B) 0.5 per cent $CuSO_4$ $5H_2O$ in 1 per cent potassium tartrate
(C) Alkaline copper sulphate solution prepared by mixing (A) and (B) in the ratio 50 : 1
(D) Folin and Ciocalteu's reagent (BDH) mixed with distilled water in the ratio 1 : 2.

A standard curve is first prepared using bovine serum albumin (Armour Laboratories). From a stock solution containing 20 mg/ml in normal saline, dilutions are prepared so as to give solutions containing 17.5 : 15 : 12.5 : 10 : 7.5 : 5 : 2.5 and 1 mg/ml. To each of ten test tubes are added 5 ml of reagent (C) and then 0.02 ml aliquots of each of the nine protein solutions (Drummond disposable micropipettes are a suitable means of measuring this small volume). The solutions are gently shaken to ensure mixing. A 0.02 ml aliquot of normal saline is added to the remaining tube to serve as a blank. Ten minutes later, 0.5 ml of reagent (D) is added to each tube and the contents mixed. The solutions will be seen to turn varying shades of blue depending upon the concentration of protein present in the aliquot; this blue colour is allowed to develop for 20 minutes. The intensity of colouration for each of the tubes is then measured against the blank in a spectrophotometer at a wavelength of 600 nm. Optical density readings are plotted against protein concentration to produce a standard curve. To determine the concentration of the unknown protein solution, a 0.02 ml aliquot is taken, the colour developed, and the optical density measured against a blank, in exactly the same was as described for producing each of the values on the standard curve. The concentration of the unknown solution is then determined from the standard curve.

Preparation of antiserum

0.5 ml of the solution of crystallins (containing about 5 mg of protein), is mixed with 0.5 ml of Complete Freund's Adjuvant (Difco Laboratories). The mixture is thoroughly emulsified by continuously drawing it in, and passing it out, through a 19 guage 2in disposable needle attached to a disposable 1 ml syringe. The stability of the emulsion can be tested by adding a small drop to water contained in a beaker; no dispersion of the drop should occur. The purpose of the adjuvant is to retain the protein at the injection site and provoke infiltration of macrophages and lymphocytes. A greater antibody response is achieved in this way than by simply injecting the protein solution directly into the blood, since from here it would be more quickly eliminated. An adult rabbit is injected subcutaneously with the emulsion, in three or four sites in the nape of the neck. Similar injections are repeated once a week for three weeks; the rabbit is left for a month and a further injection is then given. Blood is obtained from the marginal ear vein and from the heart, after the rabbit has been killed with an intravenous injection of nembutal (sodium pentobarbitone – Abbot Laboratories). The blood is allowed to clot and the serum separated by centrifugation. Such antiserum can conveniently be stored in 5 ml aliquots, in bijoux bottles, at $-20°C$. In this condition it will retain its potency for a number of years.

A simple test for the presence of precipitating antibodies (the so-called 'ring test'), is to bring the antiserum and the protein antigens into contact in a narrow tube. To do this a short piece of clean 2 mm internal bore glass tubing is drawn out to form a micropipette, and inserted into a short piece of tubing attached to a syringe (as depicted in Fig. 6.16). A small volume of lens proteins is drawn into the tube and then a similar volume of antiserum, so that a sharp interface is maintained. The end of the pipette is then sealed in a bunsen flame. A ring of white precipitate will quickly form at the interface if antibodies are present in strength.

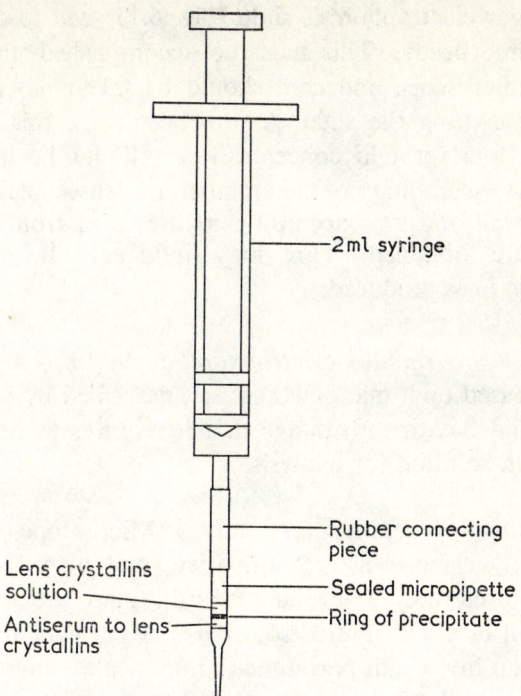

Fig. 6.16. Method of carrying out the 'ring test'.

Preparation of larval lens crystallins

Tadpoles at varying stages of development (for example stages 40, 45, 50 and 60) are killed with MS222 or by decapitation, and under the dissecting microscope the lenses extracted from corneal incisions made in the eyes. Lenses are freed from any adherent material and pooled for each appropriate stage in saline contained in a solid watch glass. They are then transferred by a micropipette to a small volume of saline (0.2 ml) contained in a micro-homogenizing tube, and homogenized. After centrifuging to remove insoluble protein, the concentration of crystallins in each of the solutions is determined as previously described, and values equated by adjusting the volumes of each of the solutions. A value of at least 5 mg/ml should be aimed for. An alternative procedure is to transfer lenses to the sample well of the agar electrophoresis slide (Fig. 6.17) and to crush them with a fine needle. This must be accomplished under a dissecting microscope and care should be taken not to damage the agar coating the slide. A drawback with this method is that the total protein concentration will not be known and may vary according to the number of lenses placed in the sample well and also according to the stage from which the lenses are obtained. This may influence the pattern of precipitin lines produced.

Immunoelectrophoretic analysis

This is based on a micromethod first described by Scheidegger (1955) and has the advantage that it requires as little as 2 μl of protein solution for analysis.

Preparation of agar coated slides Microscope slides are thoroughly cleaned in alcohol, dried, and placed on a level surface (preferably a levelling board). They are then coated with 2 ml of a hot, filtered solution of 1 per cent Ionagar No. 2 made up in sodium barbitone-sodium acetate buffer, pH 8.6, ionic strength 0.05, and containing 0.15 per cent sodium azide as an antibacterial agent. The buffer is prepared as follows: sodium barbitone 5.00 g; hydrated sodium acetate 3.33 g; 0.1 N HCl 34.2 ml. After adding HCl the solution is made up to one litre with distilled water. The agar and sodium azide are dissolved in the buffer by heating to boiling point with

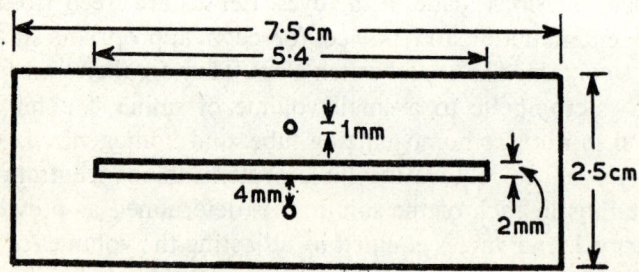

Fig. 6.17. Template for preparing micro-immunoelectrophoresis slides.

constant stirring, after which the hot solution is filtered on a Buchner flask through a Whatman GF/A glass filter paper. The agar is applied from a previously warmed pipette and should cover the slides evenly. When the agar has set the slides are placed on a template drawn to the specifications illustrated in Fig. 6.17. The agar is conveniently removed from the sample well by means of a disposable Pasteur capillary pipette (Harshaw Chemicals Limited) the narrow end of which has a diameter of 2 mm. The pipette is simply inserted over the template hole and the agar sucked out by means of an attached rubber tube. To make the lateral well, cuts are made with a scalpel blade supported against a 6 in plastic rule held just above the agar. The agar is then carefully lifted out. It is important to have very clean cuts and to have the sample wells equidistant from the lateral well. The slides should not be allowed to dry out and can be stored until required in sandwich boxes lined with filter paper saturated with water.

Electrophoresis A suitable electrophoresis tank (we use a Kohn Universal, Shandon Scientific Instrument Company) is placed on a levelling board, and the buffer and electrode compartments filled with a total of 1 l of sodium acetate-sodium barbitone buffer prepared as previously described. After the levels have been made equal in all compartments of the tank, the agar slides are placed on the supporting bridges and connected to the buffer compartments by means of Whatman 3 MM filter paper wicks that have been saturated with buffer. (Fig. 6.18 shows the final arrangement of the set up in plan view.) The sample holes are filled with appropriate solutions of lens crystallins from adult and larval lenses. This is best achieved by means of micropipettes controlled by suction from the mouth via rubber tubing. Care should be taken not to flood the sample wells since this may subsequently lead to distortion of the precipitin lines. A water seal is made around the lid of the electrophoresis tank and a constant direct current supplied from a power pack (Shandon 'Vokam'). Adequate separation is usually achieved in one and a half hours with a current of 30 mA (approximately 6v/cm, as measured between the filter paper wicks). At the end of the run the current is switched off, the wicks removed and the lateral well carefully filled with rabbit antiserum to adult *Xenopus* crystallins. The lid of the bath is then replaced and the slides left to develop precipitin lines, a process which takes about 24 hours for completion.

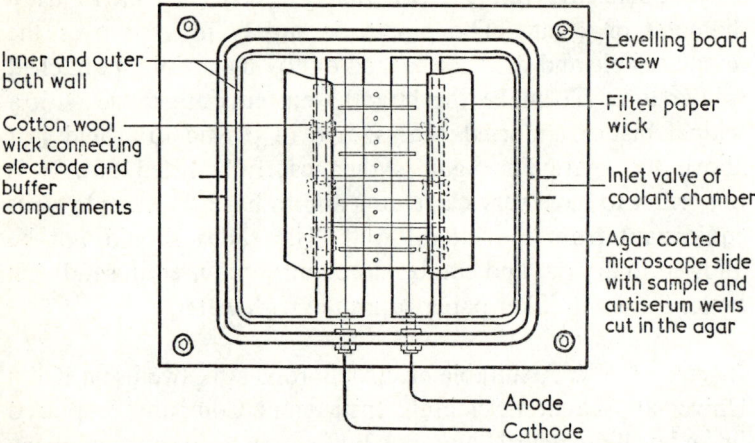

Fig. 6.18. Plan view of the electrophoresis apparatus set up for carrying out micro-immunoelectrophoresis on agar coated slides.

Staining the slides Permanent preparations can be made by drying the agar and staining the precipitin lines. The slides are first placed in 0.85 per cent saline contained in a sandwich box and left for three to four days with frequent changes of saline. This is in order to wash out the non-precipitated proteins from the agar which would otherwise obscure the precipitin lines when these are stained. The agar is then dried to a thin film on the slide by applying a piece of filter paper, in which numerous holes have been pierced with a pin, so that it completely covers the surface of the agar. Such slides are then left at 37°C to dry. The buffer salts pass into the filter paper as they dry and the pin holes cause the paper to lift off the slide rather than stick to it. The precipitin lines are then stained for 30 minutes by immersing the slides in 0.05 per cent azocarmine (made up in sodium barbitone-sodium acetate buffer, and containing 5 per cent glacial acetic acid and 10 per cent glycerine). Excess stain is washed off the slides by immersing them in 5 per cent glacial acetic acid. They are finally rinsed in distilled water and left to dry. Photographs of the precipitin lines are easily obtained by using the slides as one would a film negative. They are placed in an enlarger and the image projected onto photographic paper.

Results

The number of precipitin lines produced by the adult and larval lens crystallin solutions will vary according to the potency of the antiserum obtained, but the major precipitin lines in the γ-, β-, and α-regions should be easy to observe. Campbell, Clayton and Truman (1968) have reported detecting as many as 22 separate antigens with antiserum produced by a prolonged immunization schedule. Typical results illustrating the major precipitin lines that we have found, are depicted diagramatically in Fig. 6.19. It would appear that γ-crystallins are the major proteins present in the youngest tadpoles; β-crystallins become more evident at stage 45, and α-crystallins become evident at stage 50. The adult pattern is probably not fully developed until some time after metamorphosis. Since the antiserum has been prepared against adult *Xenopus* lens crystallins, it will not detect any crystallins that might possibly be specific to any of the tadpole stages.

It has been established (Clayton, 1970) that the different crystallins may in some cases have sub-units (polypeptide chains) in common. This is reflected in the precipitin lines by certain of them linking up and forming so-called 'cross reactions'. Clayton has suggested that the reason for the changing pattern seen during ontogeny may in some way be related to the need to maintain the correct refractive index of the lens as it grows in size.

Attempts have been made to locate the crystallins more precisely in the lens and lens-forming tissues, by means of immunofluorescence (Takata, Albright and Yamada, 1964; 1965; Campbell, 1965, Zwaan and Ikelda, 1968). In this technique (described more fully in relationship to the detection of immunoglobulin in rabbit foetal membranes, p. 227) specific antisera to α-, β-, and γ-crystallins are conjugated to a fluorescent dye, and sections of the developing eye, fixed and embedded in such a way as to preserve the antigenicity of the lens proteins, are exposed to the conjugated antisera. Subsequent exposure of the sections to u.v. light in the fluorescence microscope reveals sites of specific fluorescence indicative of the crystallins. Such studies have been carried out mainly on

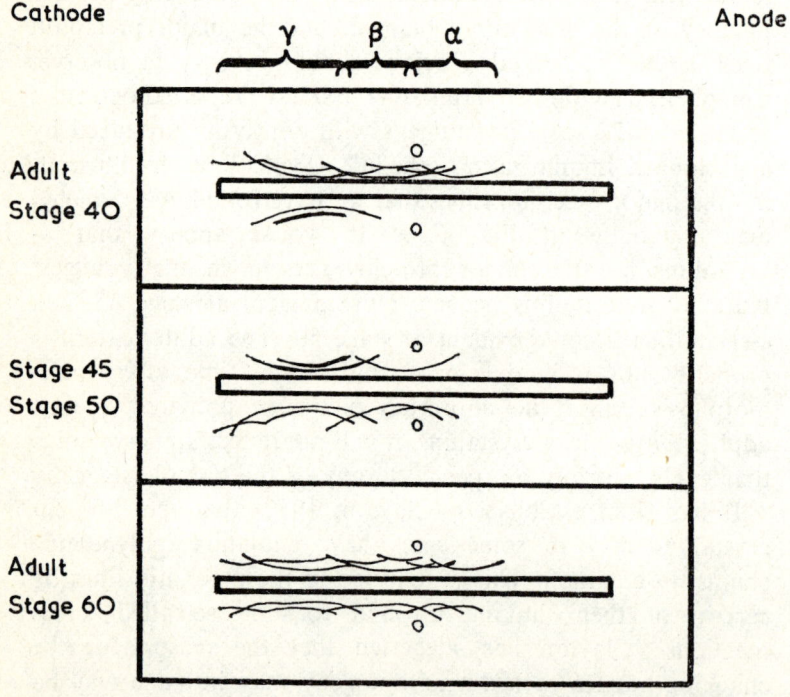

Fig. 6.19. Diagram showing typical precipitin lines produced after immunoelectrophoretic analysis of stage 40, 45, 50 and 60 larval, and adult, *Xenopus* lens crystallins. Precipitin lines were developed with rabbit anti-adult *Xenopus laevis* lens crystallins. α, β and γ, denote regions of different lens crystallin classes.

regenerating lenses, where it seems to be generally agreed that γ-crystallins are the first to appear, principally in prospective primary fibre cells. Chick lenses apparently have no crystallin equivalent to the γ-crystallin of other species, but the first to appear during differentiation has been variously designated FISC (First Important Soluble Crystallin), β-crystallin, or δ-crystallin (Zwaan and Ikelda, 1968). Early, less refined studies on lens crystallin differentiation, led to the notion that the capacity for tissues to produce a lens was somehow related to the presence of the lens antigens within them. However, lens crystallins have not been detected in normal iris tissue (Yamada, 1967) which in *Triturus* is the site of lens regeneration, and it is now more generally accepted (Zwaan, 1968) that production of lens crystallins is the result of lens cell differentiation, and not the cause.

Thyroxine induced regression of isolated Xenopus laevis tadpole tails

Amphibian metamorphosis is a process involving dramatic developmental changes, which morphologically have many manifestations, including the loss of the tadpole tail, eruption of the fore-limbs, thickening of the skin, development of the tympanum, and shedding of the horny teeth and jaw lining. Besides such morphological changes there are also distinct biochemical alterations. Nitrogen excretion for example, becomes a ureotelic rather than an ammonotelic process, and associated with this change, arginase (the enzyme responsible for catalyzing the hydrolysis of arginine to urea in the ammonia-urea cycle) rises rapidly in concentration. To meet altered osmotic demands, albumin production by the liver is also greatly increased, and there is a switch in production of haemoglobin from foetal to adult type. It has long been known that thyroid hormones, under the control of the pituitary gland, are implicated in such changes. When thyroxine or triiodothyronine are injected into young tadpoles, or merely added to their surrounding water, precocious metamorphosis may take place; thyroidectomy, on the other hand, will prevent metamorphosis from occurring (see Etkin, 1968 and Frieden, 1968, for respective reviews of the hormonal control and biochemistry of amphibian metamorphosis.)

The action of thyroxine is very specific and tissues have an inbuilt capacity to respond to particular concentrations of hormone. Some tissues are not affected and mention has already been made of the failure of eye tissues, formed from optic vesicles transplanted to the tadpole tail, to regress at metamorphosis. Such action of thyroid hormones on development has been elegantly and simply demonstrated by Weber (1962), Shaffer (1963) and Tata (1966), all of whom used *Xenopus laevis* tadpole tails cultured with or without hormone in an *in vitro* system. This technique can easily be adapted for use in the teaching laboratory and makes an interesting and rewarding individual or group study for advanced students. Practical details are as follows.

Animals

It is essential to use prometamorphic *Xenopus laevis* tadpoles (stages 50 to 55, total length 34 to 39 mm) since smaller tadpoles may give very variable results. They should be pretreated for 24 hours in aged tap water containing 0.025 per cent sulphadiazine, so as to reduce the risk of bacterial infection.

Operations

All instruments, pipettes, and dishes, should be sterilized before operations. Sterile Holtfreter's solution, which is used as culture fluid, should contain 0.05 per cent sulphadiazine. Tadpoles are immersed in Holtfreter's solution for about 15 minutes and then placed in a 1:7000 dilution of MS222 to anaesthetize them. When still, they are transferred to a fresh dish containing Holtfreter's solution and their tails (10 to 12 mm, or about one-third the whole body length) snipped off with scissors. The tails are then washed three times in fresh solution and placed in groups of three in sterile petri dishes containing 9 ml of Holtfreter's solution. The petri dishes are numbered, and before any other substances are added to the dishes, left at room temperature (18 to 20°C) for three days in order for the cut surfaces of the tails to heal. Availability of suitable embryos will dictate how many petri dishes containing tails are set up, but in order to allow for losses due to infection, five is a reasonable minimum number.

The length of the tail is measured one day after the operation, and on every subsequent day for 14 days. A millimeter scale is drawn on a white card and placed on a black background beneath the petri dish containing the tails. When illuminated, the tails show up as transparent grey structures against the black background and the scale is lined up alongside them. After three days of culture, thyroxine is added to the culture fluid so as to give a final concentration of 1 μg/ml. A stock solution is first prepared by dissolving 1 mg of sodium L thyroxine (BDH) in a small volume of 0.1N NaOH, and making this up to 100 ml with Holtfreter's solution. The pH should be adjusted to 7.2. To each experimental culture of tails is added 1 ml of thyroxine solution. Control cultures should receive the same volume of solvent and culture fluid.

Results

After discarding any obvious abnormalities due to bacterial contamination (in which case the tails will become opaque and fluffy) the average length of the tails is plotted against time. Fig. 6.20 shows the results of a typical experiment. Control tails, as well as those exposed to thyroxine, normally show a reduction in length, but by about day six of culture (three days after adding thyroxine) treated tails should start to show a definite regression. This will become more marked, compared to controls, as time progresses. Observations should be made of the microscopic appearance of the cultured tails and specimens should be selected from control and experimental groups at days 7 and 12 for fixation and histological processing. From microscopic observations of stained sections, it will be seen that changes, characteristic of normal tail regression at metamorphosis, are induced by thyroxine treatment of cultured tails. Such changes include thickening of the epi-

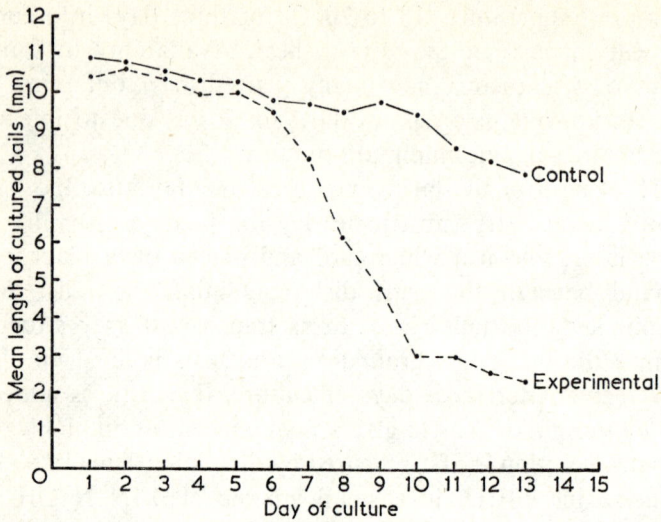

Fig. 6.20. Comparison of rates of regression of mean lengths of *Xenopus laevis* tails cultured in control under experimental conditions. Thyroxine was added to the experimental cultures on day 3.

dermis, diminution of the dorsal and ventral fins, and finally degeneration of the muscle, notochord and nerve chord. At about ten days of culture, even control tails will start to show some signs of tissue breakdown; this is probably due to the exhaustion of essential metabolites.

The fact that there is a three day lag period before differences between the experimental and control cultures become manifest, indicates that thyroxine is not having a direct effect upon the tail tissues. An interesting extension of this experiment is to add puromycin (12 μg/ml final concentration) as well as thyroxine to the culture fluid. Successful experiments will show that there is now much less difference in tail length when compared to controls. Using this type of experiment, Tata (1966) has shown that RNA and protein synthesis are essential prequisites for thyroxine induced regression of tadpole tails. Such synthesis is necessary for the production of lysosomal enzymes, which when liberated from cells, cause the involution of tail tissues (Weber, 1969).

REFERENCES

Bebbington, A. and Thompson, T. E. (1967) 'Teaching experimental embryology, I' *School Science Review*, **167**, 85–94.

Billett, F. S. (1968), 'Cellular differentiation in ectodermal explants from amphibian gastrulae', In: McGee-Russell, S. M. and Ross K. F. A. (Eds) *Cell structure and its interpretation*, Edward Arnold, London.

Billett, F. S. and Brahma, S. K. (1960), 'The effect of Benzimadazole on the differentiation of ectodermal explants from the gastrulae of *Xenopis laevis*', *J. Embryol. Exp. Morph.*, **8**, 396–404.

Campbell, J. C. (1965), 'An immuno-fluorescent study of lens regeneration in larval *Xenopus laevis*', *J. Embryol. Exp. Morph.*, **13**, 171–179.

Campbell, J. C. and Jones, K. W. (1968), 'The *in vitro* development of lens from cornea of larval *Xenopus laevis*', *Dev. Biol.* **17**, 1–15.

Campbell, J. C., Clayton, R. M. and Trueman, D. E. S. (1968), 'Antigens of the lens of *Xenopus laevis*', *Exptl. Eye Res.* **7**, 4–10.

Clayton, R. M. (1970), 'Problems of differentiation in the vertebrate lens'. *Curr. Top. Develop. Biol.*, **5**, 115–180.

Deuchar, E. M. (1966), *Biochemical aspects of amphibian development*. Methuen, London.

Etkin, W. (1968), 'Hormonal control of amphibian metamorphosis'. In Etkin, W. and Gilbert, L. I. (Eds) *Metamorphosis: A problem in Developmental Biology* pp. 313–348. Appleton-Century-Crofts, New York.

Freeman, G. (1963) 'Lens regeneration from the cornea in *Xenopus laevis*', *J. exp. Zool.*, **154**, 39–65.

Frieden, E. (1968), 'Biochemistry of amphibian metamorphosis', In: Etkin, W. and Gilbert, L. I. (Eds) *Metamorphosis: A problem in Developmental Biology*, Appleton-Century-Crofts, New York.

Grabar, P. and Burtin, P. (1964), *Immuno-electrophoretic Analysis*, Elsevier Publishing Company, Amsterdam.

Gurdon, J. (1967), 'African Clawed Frogs', In: Wilt, F. and Wessels, N. K. (Eds) *Methods in Developmental Biology*, Crowell, New York.

Hale, L. J. (1958), *Biological Laboratory Data*, Methuen, London.

Hamburger, V. (1960), *A Manual of Experimental Embryology*, pp. 211–213, University of Chicago Press.

Holtfreter, J. (1933), 'Die totale Exogastrulation, eine Selbstablosung des Ektoderms von Entomesoderm. Entwicklung und funktionelles Verhalten nervenlosen Organ', *Arch. EntwMech. Org.*, **129**, 669–793.

Horstadius, S. (1950), *The Neural Crest*, Oxford University Press, Oxford.

Humphrey, R. R. (1962), 'Mexican axolotls, dark and mutant white strains: care of experimental animals', *Bull. Phila. Herpetol. Soc.*, **10**, 21–25.

Ingram, A. J. (1969), 'Tumour Induction in the Axolotl (*Ambystoma mexicanum*)', *Thesis*, University of Southampton.

King, T. J. (1967) 'Amphibian Nuclear Transplantation', In: Wilt, F., and Wessels, N. K. (Eds) *Methods in Developmental Biology*, 737–791. Crowell, New York.

Lowry, O. H. Rosebrough, N. J. Farr, A. L. and Randall, R. J. (1951) 'Protein measurement with the Folin-phenol reagent', *J. Biol. Chem.*, **193**, 265–275.

Nieuwkoop, P. D. and Faber, J. (1967), *Normal Tables of Xenopus laevis* (*Daudin*), 2nd Ed. North-Holland, Amsterdam.

Rugh, R. (1962), *Experimental Embryology*, Burgess Publishing Company, Minneapolis.

Saxén, L. and Toivonen, S. (1962), *Primary Embryonic Induction* Logos Press/Academic Press, London.

Scheidegger, J. J. (1955), 'Une micro méthode de l'immuno-électrophorése', *Int. Arch. Allergy*, **7**, 103–110.

Schwind, J. (1933), 'Tissue specificity at the time of metamorphosis in frog larvae', *J. exp. Zool.*, **66**, 1–14.

Shaffer, B. M. (1963), 'The isolated *Xenopus laevis* tail: A preparation for studying the central nervous system and metamorphosis in culture', *J. Embryol. exp. Morph.*, **11**, 77–90.

Spemann, H. (1938), *Embryonic Development and Induction*. Reprinted in 1962 and published by Hafner Publishing Company, New York.

Spiegel, M. (1951), 'Chemical method for decapsulating amphibian embryos', *Anat. Rec.*, **111**, 544.

Takata, C. Albright, J. F. and Yamada, T. (1964), 'Lens antigens in a lens regenerating system studied by immunofluorescence', *Dev. Biol.*, **9**, 385–397.

Takata, C., Albright, J. F. and Yamada, T. (1965), 'Lens fibre differentiation and γ-crystallin: Immunoflourescent study of Wolffian regeneration', *Science, N.Y.*, **147**, 1299–1301.

Tata, J. R. (1966), 'Requirement for RNA and protein synthesis for induced regression of tadpole tails in organ culture', *Dev. Biol.*, **13**, 77–94.

Turner, S. C. (1973), 'The endocrinology of Xenopus laevis; the thyroid and pituitary and their relationships to growth and differentiation', *Thesis*, University of London.

Vogt, W. (1929), 'Gestaltungsanalyse am Amphibienkeim mit ortleicher Vitälfarbung', *Arch. Entwmech*, **120**, p 384.
Waggoner, P. W. (1973), 'Lens differentiation from the cornea following lens extirpation and cornea transplantation', *J. exp. Zool.*, **186**, 97–109.
Weber, R. (1962) 'Induced metamorphosis in isolated tails of *Xenopus laevis*, *Experientia*, **18**, 84–85.
Weber, R. (1969), 'Tissue involution and lysosomal enzymes during anuran metamorphosis', In Dingle, J. R. and Fell, H. B. (Eds), *Lysosomes in Biology and Pathology*, 2, Frontiers of Biology, Vol. **14**, North-Holland, Amsterdam.
Yamada, T. (1960), 'A chemical approach to the problem of the organizer', *Advances in Morphogenesis*, **1**, 1–50.
Yamada, T. (1962), 'The inductive phenomenon as a tool for understanding the basic mechanism of differentiation', *J. cell comp. Physiol.*, **60**, 49–64.
Yamada, T. (1967), 'Cellular and sub-cellular events in Wolffian lens regeneration', In: Moscona, A. A. and Monroy, A. (Eds), *Current Topics in Developmental Biology*, Vol. **2**, pp. 247–283, Academic Press, New York.
Zwaan, J. (1968), 'Lens specific antigens and cytodifferentiation in the developing lens', *J. Cell. Physiol Suppl.* **1**, **72**, 47–72.
Zwaan, J. and Ikeda, A. (1968), 'Macromolecular events during differentiation of the chicken eye lens', *Exp. Eye Res.*, **7**, 301–311.

7 Birds

For practical and historical reasons the study of bird development is almost synonomous with the study of the development of the chick. The practice of rearing and keeping chickens is probably as old as society itself. By the time of Pliny (23 – 79 A.D.) several methods of incubating eggs appear to have been well established, and on the scientific side, the early history of embryology is very largely bound up with the study of chick development (Fabricius 16th C; Harvey 17th C; Wolff 18th C). This intense interest in chick development extends of course to modern times, where the experimental analysis of organogenesis, especially limb development, owes much to work on this animal. Despite this wealth of information however, the usefulness of the chick as class material for large groups is rather limited. At an elementary level it is probably best suited for demonstrating the main features of the early development of vertebrate structure. In this chapter we have concentrated mainly on this aspect and give accounts of methods for examining and culturing the early chick blastoderm. Some attention is also paid to the embryonic membranes and in this connection the procedure for chorio-allantoic grafting is described. Before describing this practical work in detail, a few general points need to be made.

The best eggs to use are those which have a fairly thin white shell (e.g., White Leghorn) which have been obtained from an established breeder, preferably one experienced in the supply of material for teaching and research purposes. The fertility of the eggs varies during the course of the year, being higher in the spring than in the summer, and this must be borne in mind when ordering them. Although dealers will supply eggs incubated to different stages, it is usually more convenient, from the point of view of timing, to obtain recently laid eggs and to incubate them to the required stages in the laboratory. During the incubation the eggs should be placed with their longitudinal axis horizontal, and if the incubation lasts longer than 48 hours it is advisable to turn them gently from side to side, by hand, once every 24 hours or so. The incubation is carried out in a humid atmosphere at about 38.5°C. Fertilized eggs which have just been laid will be at early cleavage stages and, depending largely on the temperature, will reach head fold or early somite stages after 24 hours incubation. Eggs which are not required immediately must be stored in a cool place. For this purpose we recommend a cool humid chamber running between a temperature of 12 and 15°C; under these conditions eggs can be held for at least a week without a noticeable increase of abnormal development.

Examination of the Early Chick Blastoderm

The following account concentrates mainly on the blastoderm which has been incubated to a stage of about 15 to 18 somites, that is to the point just before the head of the embryo turns to one side. Several of the main features of early vertebrate development can be seen very easily in a fresh preparation at this stage and it therefore provides a very good starting point for the study of chick development. Once the basic structure of the 15 to 18 somite embryo has been grasped, both the preceding and succeeding stages are easier to understand. It also becomes a relatively easy matter to relate stained sections of embryos to the appearance of intact specimens.

Under normal incubating conditions the embryo will begin to turn onto its right side between about 40 and 48 hours. To obtain the 15 to 18 somite stage it is best to incubate a batch of 12 or more eggs to cover the period indicated; that is incubate one-third of the eggs for 40 hours, one-third for 44 hours and one-third for 48 hours. A staggered incubation of this kind is very convenient for a large class as it will provide a range of material which will clearly show the development of somites, the development of the primary brain vesicles and the initiation of the heart beat. The technique for handling the eggs, which is described below, is based on the initial steps of New's method for culturing the early chick blastoderm. Thus this practical may be regarded as a prerequisite for the one dealing with New's culture which is described on p. 172.

The isolation of the chick blastoderm
Holding the egg in one hand and a pair of blunt forceps in the other the egg shell is broken gently at the blunt end and removed piecemeal until about half the shell remains. If the blastoderm is located as soon as the egg is opened the chances of damaging it are lessened. Care should also be taken to avoid puncturing the vitelline membrane during this and subsequent stages. During the removal of the shell the albumen is poured into a beaker until only a little remains. The yolk is then tipped into about 500 ml of saline contained in a suitable vessel, such as a domestic pie dish. The saline should just cover the yolk and the container must allow freedom of movement for scissors and forceps. At this stage any thick albumen which adheres to the vitelline membrane should be removed with fine forceps.

The next stage involves removing the vitelline membrane from the yolk and then detaching the blastoderm from the membrane. Good illumination is essential for this operation and a black background is recommended. The membrane is removed by cutting into it around the equator of the yolk and then peeling the membrane away from the yolk, using two pairs of fine forceps. Once the membrane is free it is floated onto a watchglass with the blastoderm uppermost. The watch glass is used to transfer the vitelline membrane and attached blastoderm to a large petri dish containing saline. With the aid of a low power binocular microscope ($\times$ 10) and using a pair of fine forceps, the edge of the blastoderm is gently detached and pulled away from the membrane. Once the blastoderm is free it can be floated onto a glass square for viewing under a binocular microscope. The embryo is best seen against a black background at magnifications of times 10 to times 20, and by illuminating it from above. The procedure just described is summarized in Fig. 7.1.

As mentioned earlier, at 40 to 48 hours incubation the embryo will possess a number of features which are typical of the early stages of organogenesis in vertebrates and which can easily be recognized. Obviously, the exact stage reached will depend on the time and temperature of the incubation, but it should be relatively easy to identify somites, neural tube, brain vesicles, optic vesicles and heart. A convenient way of staging the embryos is simply to count the number of somites, and to relate this to the appearance of the other organs and to the time of incubation. The blastoderm at this stage is clearly divided into an inner clear area (*area pellucida*) in which the embryo is situated and an outer opaque area (*area opaca*) (Fig. 7.4).

The method described above can, of course, be applied to earlier and later stages of development. Blastoderms are easy to remove at the primitive streak stage (about 22 hours incubation) and onwards. The pre-streak stages present more difficulty as they frequently remain attached to the yolk when the vitelline membrane is removed. The method works well for eggs incubated for periods of up to four days. Beyond this age it is best to remove the embryo and its accompanying membranes, including the yolk sac, simply by breaking the shell, and pouring the entire contents of the egg into saline. These later embryos are more obviously alive and respond vigorously to stimuli. To avoid any possibility of suffering, these embryos should not be allowed to survive once they have been removed from the egg. They can be killed either by nicking the spinal cord in the neck region or by adding MS222 to the saline. We do not propose to give a detailed account of the stages mentioned above as they are dealt with in many text books of vertebrate embryology. For a full description, reference should be made to either Lillie (1952) or Romanoff (1960).

In any batch of eggs a small percentage will be unfertilized

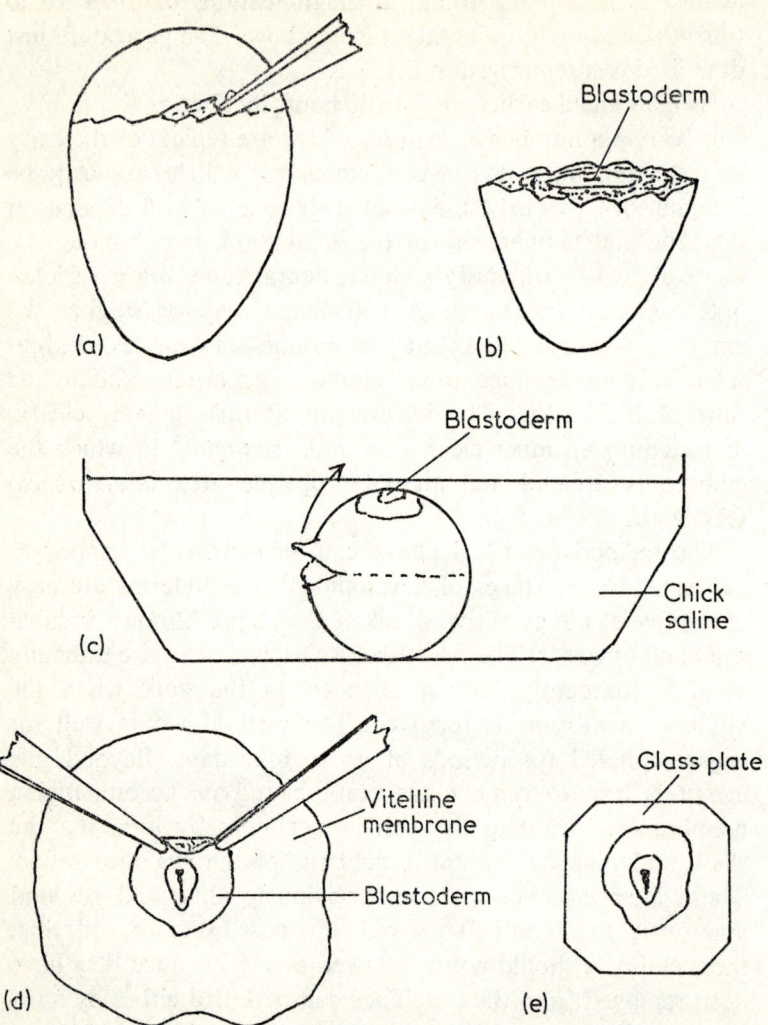

Fig. 7.1. *Removal of the chick blastoderm.* (a) The egg is opened at the blunt end. (b) About half the shell and as much albumen as possible is removed. (c) The blastoderm attached to the vitelline membrane is removed from the yolk. (d – e) The blastoderm is detached from the vitelline membrane and mounted on a glass plate.

and a small number will be abnormal. It is easy to recognize an unfertilized egg for, instead of a large blastoderm (*c.* 10 mm diam. at 40 hours), there will be a small white opaque spot (*c.* 2 to 3 mm diam.). Most abnormal embryos will only be detected after they have been removed from the egg. Abnormal embryos often appear stunted but occasionally more interesting deformities, for example, double headed embryos are seen.

Stained preparations
Although the examination of the fresh blastoderm alone is usually sufficiently rewarding, slightly more detail can be seen in a good stained preparation. An excellent procedure is that described by Mahoney (1963) using anthracene blue. For this method it is convenient to place the blastoderm on a small glass square. This is then placed in a small petri dish in which the staining is to be carried out. The schedule is given in detail below.

1. Place the blastoderm on a glass square and transfer it from the vessel containing saline to a small petri dish.
2. Fix in aqueous Bouin's solution for at least 24 hours. The fixative should be dropped gently onto the blastoderm from a pipette and enough fixative added in this way to cover the blastoderm.
3. Remove the fixative and wash in 50 per cent ethanol for about 15 minutes.
4. Place in 70 per cent ethanol saturated with lithium carbonate for at least 24 hours, that is, until the picric acid (yellow colour) is completely removed from the blastoderm.
5. Place in 50 per cent ethanol for about 30 minutes and then in distilled water for about 15 minutes.
6. Stain in an aqueous solution of 0.02 per cent anthracene blue containing 5 per cent aluminium sulphate. The blastoderm will stain slowly and progressively and should be left in the stain for at least 24 hours. No differentiation is necessary and when the blastoderm is sufficiently stained it should be transferred to water.
7. Finally the material is dehydrated and mounted in Canada

balsam according to standard procedure (i.e. pass through 50 per cent ethanol, 70 per cent ethanol, 90 per cent ethanol, absolute ethanol and xylene).

The method gives excellent results with the 40 to 48 hour blastoderm and earlier stages; the preparations are sufficiently good for detailed examination, under the light microscope, at fairly high magnification. The method also works well with the three to four day blastoderms but the thickness of these preparations precludes detailed examination at the light microscope level.

Culture of Chick Blastoderms

Although there are several well known techniques for the explantation and culture of early chick blastoderms, only one of them, that is New's method (1955), appears to be really reliable for teaching purposes aimed at demonstrating the initial stages of the development of the chick *in vitro*. It is a technique which can be carried out by the students themselves, leading almost invariably to a successful result. It can also be adapted very readily for individual project work. Essentially the technique involves using the vitelline membrane, separated from the yolk, to suspend the attached blastoderm in a suitable culture vessel, so that the albumen remains in contact with the membrane while the yolk free surface of the embryo is bathed in a small amount of culture medium. Using this technique, blastoderms may be explanted after about 22 to 24 hours incubation, that is, at a streak, head fold, or early somite stage, and cultured for a further 24 to 48 hours. Details are given below but for a full account of the method reference should be made to the publications of New (1955, 1959, 1966).

Preparation for the technique involves sterilizing the necessary instruments and glassware, and making up the culture medium, which also has to be sterilized. The following items will be required. Glassware: round pyrex soufflé dishes (or equivalent) about 20 cm in diam. and 8 cm deep; small watch glasses, about 5 cm diam; glass rings, made from 2.5 mm glass rod with an inner diam. of between 20 and 22 mm; large petri dishes, about 9 cm diam; pipettes with a terminal fine bore and knee joint. Instruments: blunt forceps, fine forceps and fine scissors. Miscellaneous: small squares of cotton gauze.

The instruments and glassware should be dry sterilized (e.g. 160°C for three hours) before use.

The recommended culture medium is Pannett-Compton saline plus glucose which should be made up from the following ingredients, and 1% glucose solution.

Stock solution A:		Stock solution B:	
NaCl	96.8 g	$NaH_2PO_4.2H_2O$	2.4 g
KCl	12.4 g	$Na_2HPO_4.12H_2O$	5.61 g
$CaCl_2.2H_2O$	16.7 g	distilled H_2O	1200 ml
$MgCl_2.6H_2O$	10.2 g		
distilled H_2O	800 ml		

Immediately before use the solutions are mixed as follows. Add to 1350 ml of 1 per cent glucose 60 ml of solution A and 90 ml of solution B and mix thoroughly. This gives a total of 1500 ml and is sufficient for between 6 and 12 eggs. The solutions must be sterilized beforehand (15 lbs for 20 minutes). It is convenient to sterilize the allotted portions of A (60 ml), B (90 ml) and the 1 per cent glucose (1350 ml) separately.

Removal and culture of the blastoderm

Suitable blastoderms may be obtained from eggs which have been incubated for 20 to 24 hours at about 38.5°C. This will give embryos at definitive streak, head process, or head fold stages (see Fig. 7.2). Before opening the eggs they should be allowed to cool to room temperature. (This helps to prevent the detachment of the blastoderm during subsequent manipulation.) The blastoderm attached to just over half the vitelline membrane is removed from the egg following the procedure already described on p. 168. Before the egg is opened, however, its blunt end should be swabbed with a piece of cotton wool, soaked in ethanol, to prevent possible infection. Also, to ensure an ample supply of membrane, it should be cut just

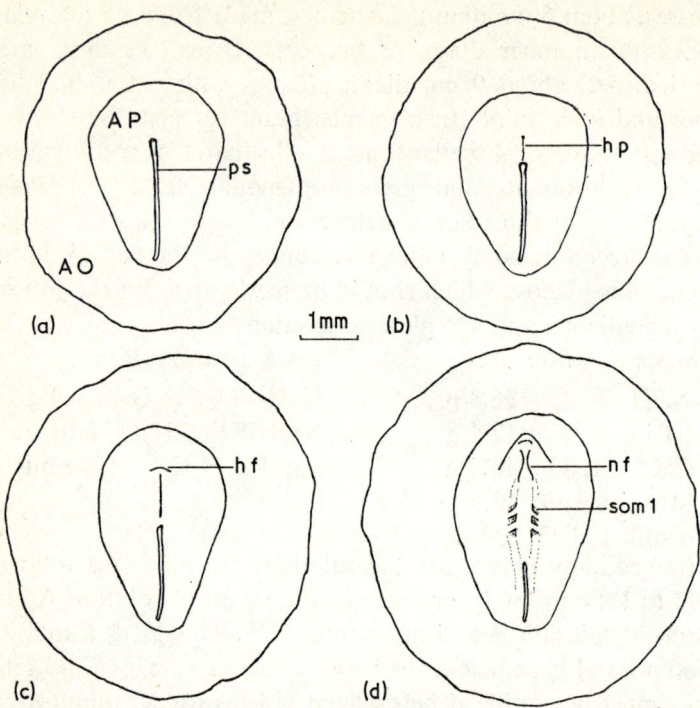

Fig. 7.2. *Chick blastoderms at an early stage of development.* (a) Full streak (Stage 4, 19 – 20 h). (b) Head process (Stage 5, 20 – 22 h). (c) Head fold (Stage 6, 22 – 24 h). (d) 2 – 3 somites (Stage 7 – 8, 24 – 26 h). *Abbreviations*: AO – area opaca; AP – area pellucida; hf – head fold; hp – head process; nf – neural fold; ps – primitive streak; som 1 – first somite. The times in parenthesis are approximate, and correspond to an incubation temperature of 38.5°C. The stage numbers are those of Hamburger and Hamilton (1951).

below the equator. As much thin albumen as possible is then collected from each egg at this stage and stored in a small stoppered conical flask. When the membrane has been freed from the yolk it is laid on a watch glass submerged in the Pannett-Compton solution, so that the blastoderm is uppermost. A glass ring is then lowered onto the vitelline membrane in such a way that the blastoderm lies in the centre of the ring and the free edge of the membrane lies outside the ring. At this stage the watch glass and its contents are lifted gently from the dish and excess saline removed from the surface by means of a fine pipette. The free edge of the vitelline membrane is then wrapped around the edge of the ring so that the membrane is stretched fairly firmly across the ring. If the membrane is stretched too tightly it may tear and if handled too roughly the blastoderm may become detached. Any tear in the membrane which allows albumen to seep through and come into contact with the upper surface of the blastoderm will cause abnormal development. A partially detached blastoderm will usually reattach itself to the membrane during culture and development will be more or less normal, depending on the degree of detachment. If the blastoderm is completely detached it is better to discard the preparation and not to waste time hoping that it will reattach and develop normally.

Once the vitelline membrane has been placed around the edge of the ring, as much saline as possible should be removed from above and below the membrane. For this purpose the fine, bent pipettes should be used. As soon as the saline has been removed about 1 ml of thin albumen is run under the membrane. At this juncture, any excess membrane may be trimmed back to the inner side of the ring. Next, the surface of the blastoderm should be cleaned of excess yolk by washing it with a small amount of saline, using the fine pipettes. It is convenient to stage the embryo during the washing procedure. When the cleaning up process has been completed, several drops of clean Pannett-Compton solution are added to the surface of the blastoderm to ensure that it remains moist during the subsequent incubation. The operation is completed by placing the watch glass in a simple moist chamber, formed by the large petri dish containing several pieces of cotton gauze which have been soaked in the saline. There is no need to incubate the blastoderms immediately after explantation as they can be left for several hours at room temperature without harm. However, if they are left for more than about half an hour it is usually necessary to add more saline to the surface of the blastoderm. The essential steps in the procedure are illustrated in Fig. 7.3.

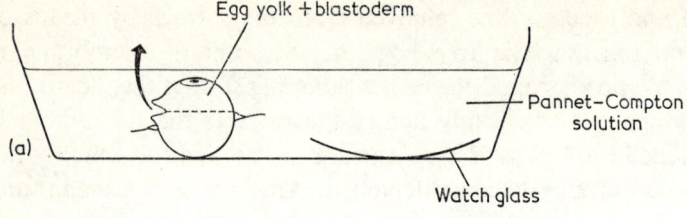
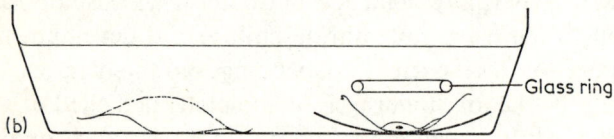
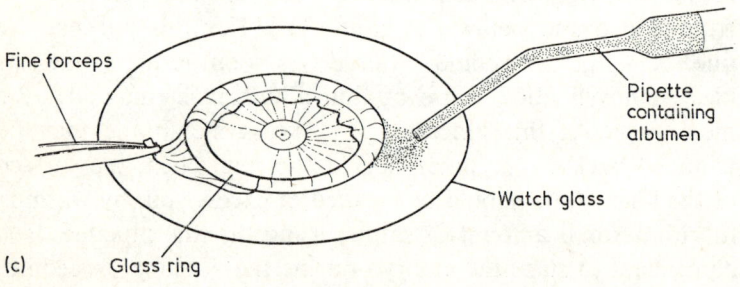
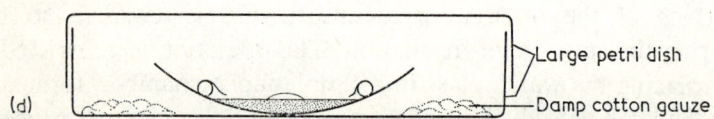

Fig. 7.3. *New's culture method*. (a) Removal of blastoderm attached to the vitelline membrane. (b) Glass ring placed over the vitelline membrane on watch glass. (c) Edge of membrane wrapped around ring. (d) Complete assembly ready for incubation.

As soon as a sufficient number of explants have been prepared, they are placed in a suitable incubator. Under normal incubating conditions, i.e. 38.5°C and a humid atmosphere, good results can be expected in the majority of cases after 24 hours. The exact stage reached after this time will depend on the starting point of the culture, but if explants of primitive streak to head fold stages were made, then embryos similar to those described on p. 169 will be obtained. That is, they will possess between 15 and 20 somites and the later stages will show a vigorous blood circulation. As a practical demonstration this is really all that is required. A typical result is illustrated in Fig. 7.4. Incubation beyond 24 hours leads to increas-

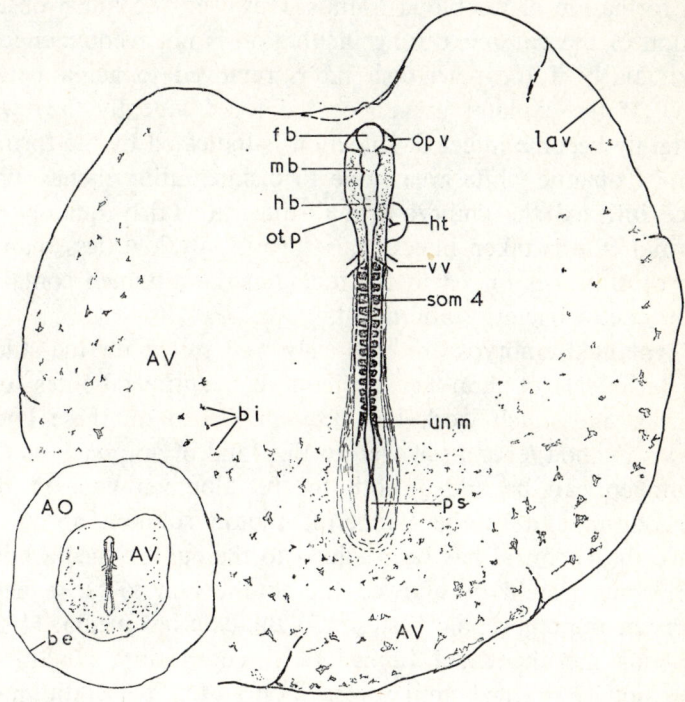

Fig. 7.4. *An explanted chick embryo.* Drawing of a 15 somite chick embryo (Hamburger and Hamilton, stage 11+) explanted at the head process stage and cultured for 24 hours, using New's technique. *Inset*: Indicates extent of blastoderm outgrowth, which is less than it would be *in vivo*. *Abbreviations*: AV – area vasculosa; AO – area opaca; be – edge of blastoderm; bi – blood islands; fb. – fore-brain; hb. – hind brain; ht – heart; lav – limit of area vasculosa; mb – mid-brain; op v – optic vesicle; ot p – otic pit; ps – primitive streak; som 4 – 4th somite; un m – unsegmented, paraxial, mesoderm; vv – vitelline vein.

ing mortality and the explanted embryos rarely last beyond 60 hours.

New's method appears to produce embryos which are very similar to those grown to the same stage *in ovo*. The only striking difference is that the outgrowth of the extra embryonic blastoderm is much less in culture than it is in the egg. A less obvious defect is the poor development of the head fold of the amnion. The method is excellent for observing the development of the heart, somites and head region, and also for studying the formation of the blood islands. However, too much observation of the cultures during incubation is not recommended, particularly if the petri dish lid is removed to get a better view! If the explants are uncovered too frequently they will certainly become infected. Infection is indicated by the formation of opaque white areas, due to disintegrating tissues, and once infected the embryo will disintegrate fairly quickly. If normal care is taken infection is surprisingly low (less than 5 per cent) and is linked to the fact that the albumen contains lysozyme, a bacteriostatic agent.

Explanted embryos are obviously well suited for the study of the effects of teratogenic agents, e.g. antimetabolites and X-rays, and much work has been published on these lines. New's method lends itself well to this kind of approach as the teratogen can be added both to the albumen beneath the blastoderm and to the Pannett-Compton solution above it. Once the chemical has been added to the culture media however, time should be allowed for equilibrium to be reached between the embryonic tissues and the external media. If the explants are incubated immediately, equilibrium conditions may not be reached until several hours after the addition of the teratogen. Thus, treatment applied at the primitive streak stage may only be effective at the head fold or early somite stage. To be on the safe side, treated explants should be left in a cold, humid chamber (*c*. 15°C) overnight (say 16 hours), before incubating at normal temperature (Billett, Collini and Hamilton, 1965). Apart from the study of the effects of teratogens explanted blastoderms can also be used for the study

BIRDS

of blastoderm outgrowth (New, 1959), the formation of blood islands (O'Brien, 1961) and the study of cell movement. In the latter case, however, Spratt's method which involves the culture of the isolated blastoderm on a solid medium is probably better as it allows a dorsal view of the embryo (Spratt, 1948).

New's method can obviously be adapted for the culture of other embryos. In the case of small eggs, such as those of quails, the amount of vitelline membrane which can be used for suspending the blastoderm is much smaller, and the culture chamber needs to be modified to take account of this. One such modification is illustrated in Fig. 7.5 and its use is described elsewhere (Billett, Bowman and Pugh, 1971). In this method New's glass ring is replaced by a ring of stainless steel

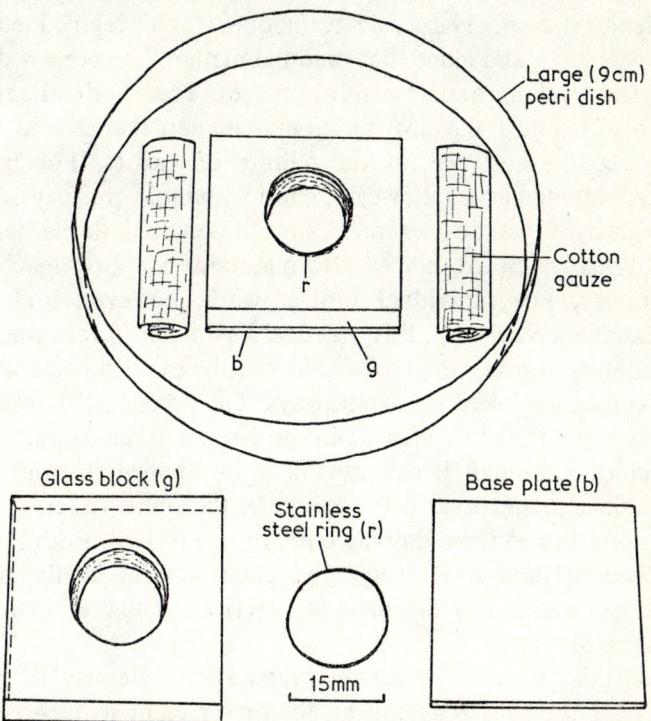

Fig. 7.5. Modified New Culture chamber used for explanted quail embryos (Billett, Bowman and Pugh, 1971).

wire (about 1 mm diameter). The wire ring rests on a ledge at the top of a central hole cut into a glass block about 5 mm thick. A glass plate held in position by silicone grease, forms the base of the central chamber which holds about 1 ml of albumen. The blastoderm is held in the assembly in the same way as in the New culture, and as in the original method a closed petri dish, containing moist gauze, is used to provide a humid chamber.

Preparation of Chorio-allantoic Grafts

The chorio-allantois of the eight to nine day old chick forms a vascular bed on which the independent development of organ primordia, isolated from younger chick embryos, can be demonstrated. The basic technique was developed some fifty years ago and since that time the method has been widely used for teaching and research purposes. The chorio-allantois will also support the growth of mammalian tissues and has been used extensively for the culture of viruses. For large undergraduate classes, however, chorio-allantoic grafting is not a satisfactory practical exercise, since it demands sterile bench conditions which are not easy to maintain in a busy teaching laboratory. For individual project work, however, and for small advanced classes, this practical has much to recommend it. It demands a fair degree of skill and the experimental situation can be exploited in several ways. The procedure described below is for the graft of a limb bud from a three to four day old chick, although it can obviously be applied to grafts of other developing organs, for example, the optic vesicles, and heart of a two to three day old embryo (see p. 169). Both Rugh (1962) and Hamburger (1960) give good accounts of the standard procedure, and the method described below is, in most respects, similar.

For this practical adequate preparation is essential. The host eggs are placed in the incubator for eight to nine days, and the donors for three to four days, before the grafts are made. The day before the practical all instruments, glassware, and solutions are sterilized. It also helps if the eggs are candled on the previous day so that time is not wasted opening infertile eggs. It is best for students to work in pairs for this practical, one being responsible for the preparation of the donor grafts, and the other for the preparation of the host embryos. Two eggs serve for the donors, and four for the hosts. It is advisable for students to wear sterile surgical face masks for this practical, to lessen the risk of contaminating the embryos.

Isolation of limb buds
After cleaning the surface of the shell of the donor with 70 per cent ethanol, the egg should be opened to expose the embryo. This can be done either by cutting a large window in the side of the shell, or by removing most of its blunt end. When the embryo has been located, it is freed from the yolk and the outer part of the blastoderm, by cutting into the blastoderm around the embryo. Immediately the embryo is free, it is removed from the egg with a large pipette or forceps and placed in sterile saline contained in a large petri dish. The limb buds should be located on the embryo under a dissecting microscope. The hind limb buds are the easiest to remove; this is done by chopping off the back end of the body just above the limb buds, and then removing the buds and some of the nearby body wall, from the separated hind quarters. Both hind limb buds are prepared in this way, and transferred to a small amount of sterile saline, contained in a watch glass, until the host is ready.

Preparation of the host embryos
The aim of the procedure is to expose a small area of the chorio-allantois to receive the graft. It is advisable to pierce the blunt end of the shell with a needle before the host egg is opened, in order to ensure the collapse of the air space when the egg is opened. If this is not done, the chorio-allantois remains near the surface; this not only increases the risk of cutting the blood vessels in the membranes, but also means the grafted tissue will develop in cramped conditions. The only

advantage of *not* collapsing the air space is that the graft is easier to find at the later stage. To expose the chorio-allantois a window (about 1 cm square) is cut into the side of the shell which has been sterilized previously with a 70 per cent ethanol. This can be done with a piece of a small hacksaw blade, but by whatever means, the cut must be made carefully. Ideally the cut is made so that it does not go through the egg membrane. When the initial cuts into the shell have been made the area is moistened with sterile saline, and then the shell membrane itself is cut into on three sides, using small scissors. The membrane is left intact on the fourth side to form a hinge for the separated piece of shell. A light is shone into the opened egg and the exposed membrane examined quickly for a place where small blood vessels form a junction, since such a place will form a suitable graft site. The donor organ, in a small drop of saline, is then transferred carefully to the host and placed gently onto the chorio-allantois. Care should be taken not to use too much fluid for the transfer. Once the graft is in position the shell of the host is closed by sealing the hinged piece of shell back into position with wax, or with suitable sterile tape. The egg is then returned to the incubator. The procedure for the preparation of chorio-allantoic grafts is summarized in Fig. 7.6.

After nine to ten days incubation the graft should be located and examined. At first an attempt should be made to find the graft by re-opening the original window and enlarging it, so that the area in which the graft was originally placed can be seen. If the graft is visible it should be carefully removed with fine scissors and forceps and placed in a drop of saline. If the graft cannot be found at the first attempt the shell should be completely removed and the host embryo placed in a dish of saline, so as to allow a thorough search of the chorio-allantoic membrane. The grafts are not always easy to detect, and any reddish nodular area should be removed for further examination. In connection with the recovery of the graft it must be remembered that the host at this stage will be near to hatching. To avoid any possibility of cruelty the young animal

BIRDS

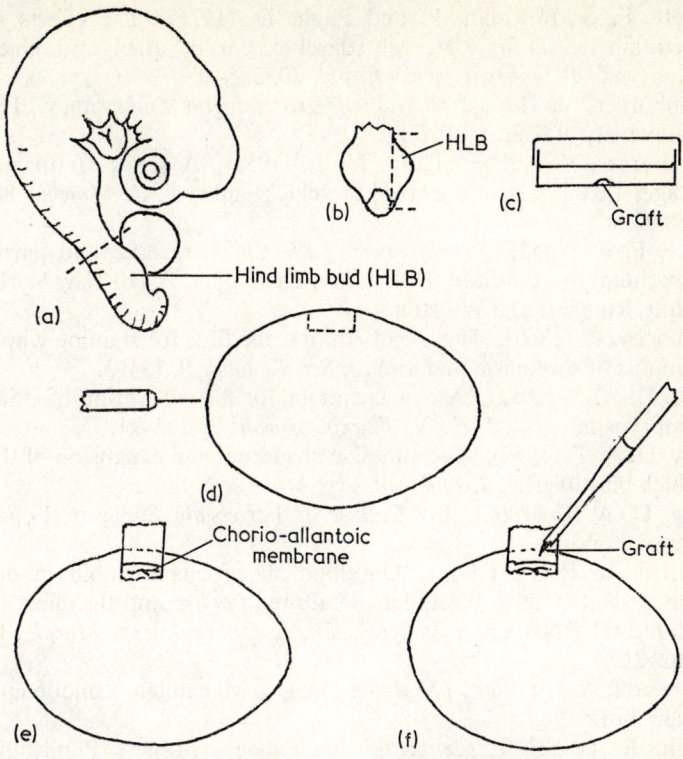

Fig. 7.6. *Chorio-allantoic grafting.* (a – c) Preparation of graft (hind limb bud). (d – f) Preparation of host and placing the graft.

should be killed immediately by cutting the spinal cord in the neck region. Although a superficial examination of the graft in saline will reveal its general morphology, the details of its structure (if it is a limb) can only be observed after using a technique which stains cartilage.

REFERENCES

Billett, F. S., Collini, R. and Hamilton, L. (1965), 'The effects of D- and L-threo-chloramphenicol on the early development of the chick embryo'. *J. Embryol. exp. Morph.*, **13**, 341–356.

Billett, F. S., Bowman, P. and Pugh, D. (1971), 'The effects of actinomycin D on the early development of quail and chick embryos', *J. Embryol. exp. Morph.*, **25**, 385–403.

Hamburger, V. (1960), *A Manual Experimental Embryology*, The University of Chicago Press.

Hamburger, V. and Hamilton, M. L. (1951), 'A series of normal stages in the development of the chick embryo', *J. Morph.*, **88**, 49–92.

Lillie, F. R. (1952), *Development of the Chick*, revised (and partly rewritten) by Hamilton, H. L. (Adv. Ed. Willier, B. H.) New York: Holt, Rinehart and Winston.

Mahoney, R. (1963), 'The use of Anthracene Blue for staining whole mounts of zoological material', *J. Sci. Technol.*, **9**, 154–5.

New, D. A. T. (1955), 'A new technique for the cultivation of chick embryos *in vitro*', *J. Embryol. exp. Morph.*, **3**, 202–21.

New, D. A. T. (1959), 'The adhesive properties and expansion of the chick blastoderm', *J. Embryol. exp. Morph.*, **7**, 146–64.

New, D. A. T. (1966), *The Culture of Vertebrate Embryos*, Logos Press, London.

O'Brien, B. R. A. (1961), 'Development of haemoglobin in de-embryonated chick blastoderms cultured *in vitro* and the effect of abnormal RNA upon its synthesis', *J. Embryol. exp. Morph.*, **8**, 202–21.

Romanoff, A. L. (1960), *The Avian Embryo*, Macmillan, London and New York.

Rugh, R. (1962), *Experimental Embryology*, Burgess Publishing Company, Minneapolis.

Spratt, N. T. (1948), 'Development of the early chick blastoderm on synthetic media', *J. exp. Zool.*, **107**, 39–64.

8 Mammals

As far as mammalian development is concerned, the choice of student practical work that is not confined simply to examination of stained sections, is limited. This is because of the inherent difficulty of observing development of mammalian embryos within the female reproductive tract, and the restrictions on experimental work imposed by the Cruelty to Animals Act. However, we have outlined details of six practical exercises which, when considered as a whole, in our opinion give students the opportunity to acquire practical skills and at the same time gain an insight into the way mammals are adapted to intra-uterine development. Details relevant to the hormonal control of reproduction, the anatomy of the male and female reproductive systems, and the appearance of the male and female gametes, are included in the first three exercises. These are relatively simple and suitable for an introductory course The other three exercises require more expertise and an extended period of practical time. Apart from the need to have some microscopes fitted with phase contrast, the only sophisticated apparatus required is a fluorescence microscope. This is required for the detection of immunoglobulin in the rabbit foetal membranes as described on p. 227, and can be used to advantage when examing spermatozoa and early embryos. Fluorescence microscopes are now standard equipment in most University Biology Departments. Modern instruments have built-in filter systems making them easy for students, under supervision, to set up and use.

Despite its small size, the laboratory mouse is in many ways the most suitable animal for studying mammalian development. Mice are cheap to buy and keep, relative to other mammals, and are very easy for students to handle. Most importantly from the point of view of organization of practical classes they breed all the year round. It is not too difficult to obtain from mice, living embryos at stages prior to implantation and we have outlined the procedures required to do this. In addition we have described how attempts can be made to culture these early stages. For examination of developmental stages after implantaton it is best to look at sections. These can be prepared by removing the uterus at appropriate times after coitus, pinning it out on wax and fixing it in aqueous Bouin's or Zenker solutions. After dehydration, the uteri are cleared in cedar wood oil (a procedure which may take several weeks) and then vacuum embedded. Sections are cut in the usual way on a microtome and stained with haematoxylin and eosin. Details concerning the reproduction and early embryology of the mouse are well described in articles by Bronson, Dagg and Snell (1966) and by Snell and Stevens (1966) in 'Biology of the Laboratory Mouse'. A particularly useful series of projected annotated drawings depicting sections through the mouse conceptus (the embryo and associated embryonic membranes) at various stages of development (four to nine days), are illustrated in the article by Snell and Stevens. Used in conjunction with corresponding stained sections, these illustrations make an invaluable teaching aid for the student.

Mammals differ considerably in the manner of formation of the foetal membranes and in the structure of the chorioallantoic placenta and such differences should be brought to the attention of students. Good comparative accounts of foetal membrane development and of structure of the chorio-allantoic placenta, are given by Mossman (1937) and by Amoroso (1952). Details of the final arrangement of the foetal membranes are much easier to see in the rabbit and we have included one practical exercise that makes use of this species and demonstrates the physiological importance of the foetal membranes. An excellent series of standard 8 mm loop films for use in a loop film projector are available on early human development (Macmillan and Co. Ltd., 4 Little Essex Street,

London, W.C.2). They consist of five cassettes entitled: Growth of the oocyte and development of the ovarian follicle; Ovulation; Fertilization; Cleavage and formation of the blastocyst; and Implantation. Well documented booklets are provided with each loop film. The films can be used to complement the practical exercises on early mouse development.

The Oestrous Cycle in the Mouse – Determination of Stages by Examination of Vaginal Smears

In female mammals, reproductive processes are characterized by cyclic changes in the histology of the reproductive tract and in sexual receptivity. Whilst such changes are under the direct control of hormones secreted by the ovary, they do in fact reflect an inbuilt periodicity in the activity of the hypothalamus relating to its control of gonadotrophin secretion by the pituitary. Such an interplay of periodic hormonal secretion has the function of ensuring successful fertilization. The recurrent period of receptivity or 'heat' is called oestrus, from which the cycle derives its name and this is the time when, in most mammals, ovulation also occurs. The cycle has been most extensively studied in laboratory rodents (see review by Bronson, Dagg and Snell, 1966); thus post-pubertal rats and mice kept separate from males are known to repeat the cycle, under laboratory conditions, at intervals of about four to five days unless subjected to pregnancy or pseudopregnancy (the result of a sterile mating). In the wild, rats and mice probably suspend the cycle for a period (termed anoestrus) during the winter; they are examples of what are nown as 'polyoestrus' mammals. Monoestrous forms (foxes, boars, wolves, in fact most wild mammals) complete a single cycle annually. In the domestic dog and cat, the oestrous cycle is repeated twice each year.

In its simplest form, the oestrous cycle in the rat and mouse can be divided into four stages; proestrus, oestrus, metoestrus and dioestrus. In the first two stages there is active growth in parts of the reproductive tract, in the next, degenerative changes and then finally a somewhat quiescent or slow growth state. There is no clear cut division between the end of one stage and the beginning of the next. The cycle can be very easily followed by examining vaginal smears, since oestrogen, the principal hormone involved in rats and mice, causes specific and predictable changes in the structure of the vaginal epithelium; less obvious changes occur in the histology of the uterus. These changes in the histology of the reproductive tract, correlated with ovarian events, are briefly outlined below.

Proestrus

This stage lasts for about 18 hours during which time the follicles in the ovary, which contain the ova, are enlarging rapidly. They produce an increasing quantity of oestrogen, which has anabolic effects on the reproductive tract. Thus the uterus becomes more vascular and distended and its contractility more pronounced. Mitoses increase in the uterine epithelium which as a consequence becomes higher, and there is a marked decrease in the number of polymorphonuclear leucocytes to be seen amongst epithelial cells; the endometrial glands hypertrophy. In the vagina, the epithelium becomes thickened (10 to 13 layers of cells) and leucocytes cease to migrate through the epithelium to the lumen. A vaginal smear will reveal small, round, nucleated epithelial cells with no, or very few, accompanying leucocytes.

Oestrus

This stage lasts for about 25 hours although the female is receptive for less than 18 hours. Ovulation occurs spontaneously at about 10 hours after the onset of oestrus and usually at night. The changes occurring in the uterus at proestrus continue into oestrus. The skin around the vaginal orifice becomes swollen and the outer layer of epithelial cells, which are typically cornified, slough off into the lumen. In early oestrus such cells may retain their nuclei, which on staining appear pyknotic, but at later stages no nuclei are visible. The vaginal smear contains hundreds of large cornified epithelial cells, or 'squames', with degenerate nuclei. Towards the end of oestrus such cells are found adhering in sheets, giving a 'cheesy' consistency to the smear.

Metoestrus

This stage is of short duration, lasting for only about 8 hours. Small follicles, and many corpora lutea formed from the ruptured follicles, are present in the ovary. The uterus shows a decrease in size and vascularity and the uterine epithelium shows signs of vacuolar degeneration; leucocytes can be observed penetrating between epithelial cells. In the vagina, what were the inner layers of the oestrous vaginal epithelium now line the lumen, the outer cornified layers having been sloughed off. Leucocytes appear in the stromal tissue and migrate through the epithelium into the lumen. Vaginal smears thus show many polymorphonuclear leucocytes, cornified epithelial cells, and occasional nucleated epithelial cells.

Dioestrus

In this stage, which lasts for about 55 hours, the ovary contains large corpora lutea from the previous ovulations. In fact, corpora lutea are present in the ovary at all stages of the cycle, persisting from follicles that have shed their ova as many as three cycles previously. However, such corpora lutea are largely inactive in the secretion of progesterone, at least in so far as any effect on the reproductive tract is concerned. It is only if pregnancy or pseudopregnancy ensues, that the corpora lutea actively synthesize progesterone. In this respect rats and mice are termed 'short cycle' mammals and differ from 'long cycle' mammals such as the guinea pig, dog and domestic ungulate in having no prolonged luteal phase and recognizable progestational changes in the reproductive tract during the oestrous cycle. In dioestrus, in the rat and mouse, the uterus has a somewhat anaemic looking appearance, is low in motility, and has a slit-like lumen. The uterine epithelium is low and columnar in appearance, endometrial glands are collapsed, and leucocytes appear in the stroma. In the vagina, the lining epithelium is thin with four to seven cell layers; mitoses are infrequent. Leucocytes are abundant in the stroma, migrating through the epithelium into the vaginal lumen. A vaginal smear taken at this stage will therefore reveal stringy mucus in which are entangled many leucocytes and a few nucleated epithelial cells.

Preparation of vaginal smears
As a practical exercise in animal development, the oestrous cycle may appear somewhat out of place. However, it is useful in that it emphasizes the importance in mammals of hormonal regulation of reproduction and development. Since it is easier to handle and keep, the mouse is the best animal to use. Ideally single mice should be studied over a period of time, but this is frequently impossible with large classes. However, if a sufficient number of mice are sampled, all stages of the oestrous cycle can be observed by students in a single practical session. Details are as follows.

Female mice that have been kept apart from males are taken at random from the colony and placed separately in numbered mouse boxes. A probe, consisting of a piece of platinum or tungsten wire inserted into a metal or glass holder at one end, and bent into a narrow loop at the other, is heated in a bunsen flame to remove any contaminating material. This probe is used to obtain cells from the vaginal lumen. The mouse is caught by the tail and allowed to grip a suitable surface – the mouse box lid for example. The tail is then held between the index finger and thumb and bent backwards, whilst the remaining fingers rest on the back of the mouse and restrain it (Fig. 8.1). Invariably the mouse will defaecate and urinate whilst being handled. Any excretion overlying the vaginal orifice should be wiped away with a tissue. The probe is then taken up in the other hand, inserted gently into the vagina, and slowly rotated. Care should be taken to ensure that the loop is narrow enough to enter the vagina easily and that the vaginal wall is not scraped. The probe is then withdrawn and any material adhering to the loop smeared onto a

MAMMALS

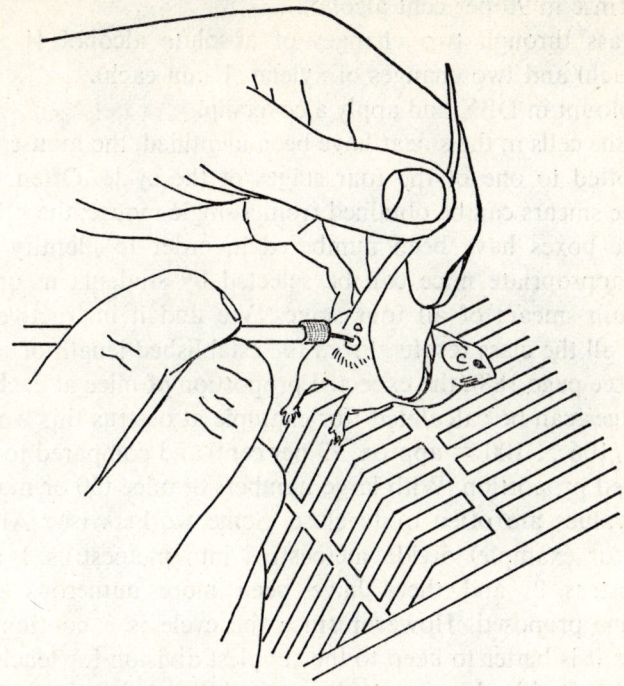

Fig. 8.1. Method of taking a vaginal smear from a mouse. The mouse is shown gripping the lid of its cage and a platinum wire loop is shown placed in the vagina.

clean microscope slide that has been numbered with a diamond. Staining of the smear with haemotoxylin and eosin is carried out as follows:
1. Allow the smear to dry on the bench and then transfer to a staining rack supported over a sink.
2. Pipette a few drops of fixative (equal parts of ether and absolute alcohol) over the smear and leave for 1 min.
3. Immerse in Ehrlich's haemtoxylin for 10 min.
4. Rinse in tap water.
5. Transfer to alkaline alcohol for 1 min.
6. Differentiate (if necessary) in acid alcohol.
7. Rinse in 90 per cent alcohol.
8. Stain in alcoholic eosin for 1 min.

9. Rinse in 90 per cent alcohol.
10. Pass through two changes of absolute alcohol (1 min each) and two changes of xylene (1 min each).
11. Mount in DPX and apply a coverslip.

When the cells in the smear have been identified, the mouse can be allotted to one of the four stages of the cycle. Often two or three smears can be obtained from a single mouse; therefore, as cage boxes have been numbered in order to identify the mice, appropriate mice can be selected by students in order to obtain smears of all four stages. We find it instructive to record all the class results. From the established length of each stage (see page 188), the expected proportion of mice at each of the stages can be calculated (for example at oestrus this would be ($25/106 \times 100 =$ approx. 23 per cent) and compared to the observed proportion. With large numbers of mice (50 or more), these values are often quite close. Some workers (see Allen, 1922, for example) divide metoestrus into metoestrus 1 and metroestrus 2, and there have been more numerous subdivisions proposed. However since the cycle is a continuous process it is better to keep to the simplest division for teaching purposes. Besides looking at the vaginal smear, it is important that students should observe sections of the ovary, uterus and vagina at different stages of the cycle, so as to familiarize themselves with the changing histological features. Such material is easy to obtain once the stage of the oestrous cycle has been determined from the smear.

With smaller groups of students and longer practical periods, some interesting extensions of the practical exercise are possible. As mentioned previously, single mice can be studied over a number of days and the length of the cycle determined. This entails taking smears at intervals of 12 hours or less. It is also possible to investigate the oestrous cycle of female mice kept together in somewhat crowded conditions. Such mice may exhibit what is known as the 'Lee Boot effect' (van der Lee and Boot, 1956) in which the normal cycle becomes suppressed and mice may go into prolonged dioestrus or pseudopregnancy. If a male is then introduced into such a crowded colony this may initiate the 'Whitten effect' (Whitten, 1957), a condition in which the cycle of the mice becomes synchronized so that a larger number than might be expected come into oestrus at the same time. By introducing male urine into the cage of suppressed females, the 'Whitten effect' can be mimicked, indicating that olfactory stimuli can modulate the oestrous cycle (Marsden and Bronson, 1964).

Examination of Unfertilized Ova from Superovulated Mice

In mammals, as in other animals, the female gametes start their development as oogonia. Within the foetal ovary the oogonia proliferate by mitotic division and become oocytes. All the eggs produced by the female mammal throughout her reproductive life are derived from oocytes that are already present at birth. The first three stages of the first meiotic division (i.e. leptotene, zygotene and pachytene) are also completed within the ovary of the foetus but at birth, or soon afterwards, meiosis is arrested in the late diplotene stage of prophase and a nucleus (the germinal vesicle) forms in the oocyte. The oocyte is now in what is commonly known as the 'dictyate' stage of development and becomes surrounded by a primary layer of follicle cells. It persists in this stage, in the ovary, until acted on by gonadotrophins secreted by the pituitary gland, and together with the enveloping cells is known as a primary follicle.

During the oestrous cycle, the growth and maturation of the primary follicles within the ovary is controlled by FSH (follicle stimulating hormone) and LH (luteinizing hormone) secreted by the anterior lobe of the pituitary. Only a limited number of follicles mature to the point where they liberate the contained ovum. Many others become atretic even though development has already commenced. It must be assumed that this is due to there being insufficient FSH and LH available. During growth of the follicles oestrogen is produced by the cells of the theca interna. As the amount produced increases, oestrogen acts both to suppress further FSH secretion by the pituitary and to stimulate LH secretion. It is the latter hormone which brings about the final maturation of the follicle and its rupture.

It has been mentioned previously that mice are spontaneous ovulators. This means that mature ova are discharged from the ovary, normally at oestrus, whether or not mating has taken place. Such ova, when liberated, come to lie in the upper part of the oviduct. This has a dilated appearance (see Fig. 8.4 and Plate 8.2a) and is called the ampulla. If mating does occur, it is here, in the ampullary region, that fertilization takes place. It is not too difficult to remove such ova from the ampulla and to examine them under the microscope, but a major problem is to be certain that ova are present, since even if the mouse can be shown to be at oestrus from a vaginal smear, there is no guarantee that ovulation will have taken place. This problem can be overcome by 'superovulating' mice a procedure which ensures that a large number of mice will have relatively large numbers of ova present in the ampulla at a more or less specified time. Large classes can thus be catered for. The technique involves injecting exogenous hormones, which have FSH and LH properties, into mice at appropriate times. This causes the gonadotrophin levels in the animals to be artificially raised, and many more follicles than is normal are induced to mature and rupture. Practical details are as follows.

Superovulation
Although adult mice can be superovulated (Fowler and Edwards, 1957) greater numbers of eggs are obtained from much younger mice (Runner and Palm, 1953). We usually work with virgin females, 6 to 10 weeks old.

The mice are injected intraperitoneally, first with 5 iu of pregnant mare's serum and then, 36 to 48 hours later, with 5 iu of human chorionic gonadotrophin. Such biological products have FSH and LH properties respectively, and are sold commercially as 'Folligon' and 'Chorulon' by Intervet Laboratories Ltd., Viking House, Bar Hill, Cambs. Ovulation occurs about 13 hours after injection of Chorulon and most

MAMMALS

eggs are found in the ampulla at about 16 hours post-injection. We find it easiest to make the injections with a 1 ml Gillette scimitar sterile disposable syringe (graduated in 100 divisions) and a 23 G × 1 needle. Although the injection can be carried out single handed, it is preferable to have two people to do it. One person grasps the loose skin behind the head between the index finger and thumb of the left hand, and holds the tail with the right hand (Fig. 8.2). The other person grasps the abdomen between the finger and thumb and having located the body wall, makes the injection into the peritoneal cavity.

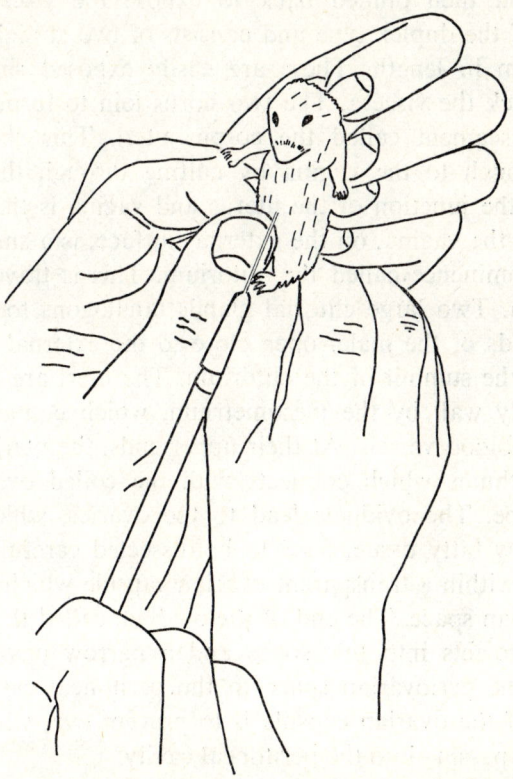

Fig. 8.2. Method of holding a mouse in order to inject it intraperitoneally.

Female reproductive system (Fig. 8.3)
Besides observing unfertilized mammalian eggs, this practical exercise affords an opportunity to examine the female reproductive system. The mice are killed for examination 13 to 16 hours after injecting Chorulon. They are placed in a killing bottle consisting of a jar of suitable capacity (the mice should not trample each other) in the bottom of which is placed a wad of cotton wool soaked in chloroform. The cotton wool is covered with a metal grill or with paper towelling so that liquid chloroform does not come into contact with the mice. The dead mouse is placed on a dissecting board, the body wall opened, and then pinned back to expose the viscera. The uterus is of the duplex type and consists of two straight horns about 4 cm in length. These are easily exposed simply by pushing back the viscera. The two horns join to form a short undivided segment called the corpus uteri. This should be traced through to the vagina by cutting through the pelvic girdle. At the junction of the uterus and vagina is the cervix. In front of the vagina, on the external surface, is a small cone shaped prominence called the clitorium. This is traversed by the urethra. Two large clitorial glands (analogous to the preputial glands of the male) open close to the external urethral orifice on the summit of the clitorium. The uteri are attached to the body wall by the mesometrium, which contains fatty tissue and blood vessels. At their upper ends, the uteri narrow into an isthmus which connects with the coiled oviduct, or uterine tube. The oviducts lead to the ovaries, which being obscured by fatty tissue, need to be dissected carefully. Each ovary lies within a transparent ovarian capsule which encloses a periovarian space. The end of the oviduct, called the infundibulum, projects into this space and a narrow passage also connects the periovarian space to the peritoneal cavity. The function of the ovarian capsule is to prevent ova, when liberated, from passing into the peritoneal cavity.

If superovulation has been successful, the ampulla can be observed with the naked eye as a more transparent, distended

MAMMALS

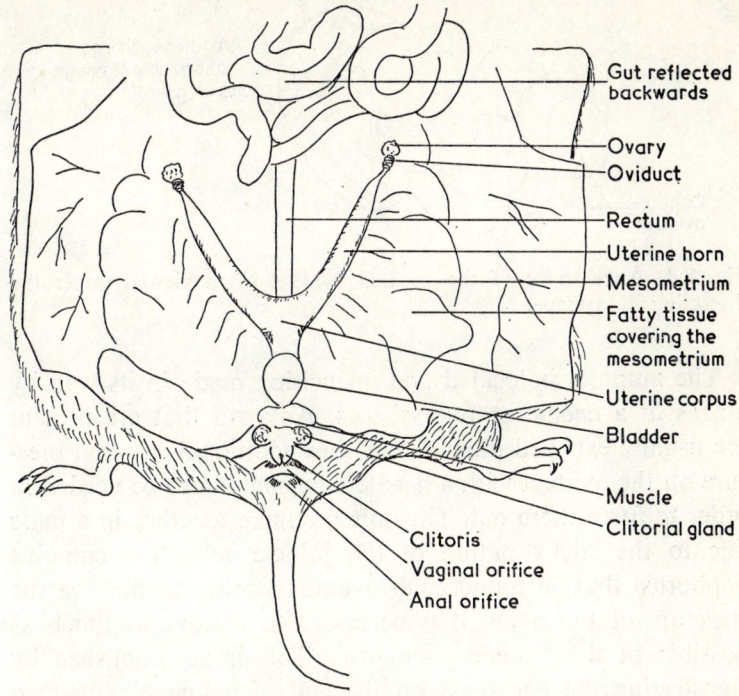

Fig. 8.3. Drawing of the partially dissected female mouse reproductive system. At this stage of the dissection, the disposition of the ovaries, oviducts and uterine horns, can be seen. The vagina has not yet been dissected in order to show its connection with the uterine corpus.

upper region of the oviduct. Distension of the oviduct is a reliable sign that ovulation has occurred since it is not observed prior to ovulation.

Isolation of living unfertilized eggs

Initially the oviduct must be dissected free by making cuts with fine scissors, first at its junction with the ovary, and then, holding the cut end with fine forceps, at its junction with the uterus. It is then carefully transferred to 0.85 per cent saline or culture fluid (see p. 215) contained in a solid watch glass (See Fig. 8.4). Subsequent procedures are carried out under a dissecting microscope using transmitted light.

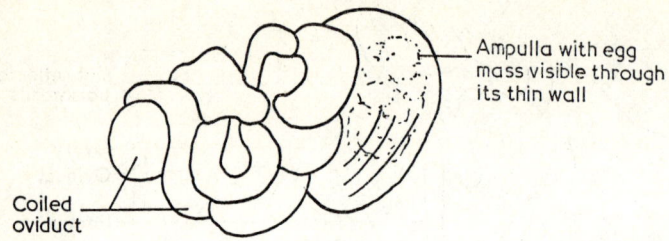

Fig. 8.4. Appearance of the isolated oviduct when viewed under the dissecting microscope.

The ampulla is located and in incision made in its wall by means of a needle or fine scissors. Any ova that are present are usually expelled immediately, but additional external pressure on the oviduct with a needle may sometimes be needed in order to force them out. They often adhere together in a mass due to the sticky nature of the follicle cells (the cumulus oophorus) that surround each ovum. In order to observe the structure of the ovum it is necessary to remove as much as possible of the cumulus oophorus. This is accomplished by transferring the egg mass on the end of a fine pipette to a solution containing 100 iu of hyaluronidase (BDH) per ml of saline or culture fluid, contained in a solid watch glass. Dispersal of the cumulus cells is aided by sucking the ova into, and blowing them out of, fine capillary pipettes. These are best made from 2 mm (internal diam.) glass tubing which should be drawn out so that the narrow part is some 3 cm long and has a bore of about 0.2 to 0.4 mm. A rubber mouth piece is attached to the other end. Care must be taken to ensure that the ova do not pass into the wide part of the pipette, otherwise they will stick to the wall and be lost. As an alternative to hyaluronidase obtained commercially, a sperm suspension can be used. This is obtained by macerating the caput and cauda epididymes of a male mouse (see p. 202) in saline. Care is needed to keep the ova under observation, otherwise they are easily lost in the cloudy fluid and may be hard to differentiate from cell debris.

When sufficient cumulus oophorus has been removed, the zona pellucida will become more easily visible. This is a transparent membrane which encloses the vitellus. It may also be possible at this stage to see the single polar body lying between the vitellus and the zona pellucida. The polar body arises from the first meiotic division of the egg which takes place just prior to ovulation. The ova are best observed however, by looking at them under the higher powers of a phase contrast microscope. To make such observations, ova are transferred by micropipette in a small volume of fluid to a microscope slide. A coverslip which has had vaseline smeared along its edges save for a small gap, is then placed on top of the fluid and pressure applied gently until the egg surface is touched. It will be appreciated that transferring ova by micropipette from solution to solution, and from solution to microscope slide, under the dissecting microscope, enables the student to gain experience in handling mammalian eggs. Such expertise is needed in order to obtain and manipulate post-fertilization mouse eggs; details of this material are described later in this chapter.

Although further development of the mammalian egg does not normally take place until after it has been fertilized, there have been some interesting recent reports that unfertilized mouse eggs obtained by superovulation can be stimulated to divide and develop further. Thus Graham (1970) has shown that by removing cumulus oophorus cells with hyaluronidase in the way described above, and transferring the ova to an appropriate culture medium, parthenogenetic development as far as the blastocyst stage can be achieved. The stimulus to further division is somehow mediated through the hyaluronidase. By applying an electric shock across the ampulla of mice that had been induced to ovulate, Tarkowski, Witkowska and Nowicka (1970) have shown that parthenogenetic development can be induced *in situ,* and that in some cases implantation of blastocysts may take place allowing normal development even as far as the eight somite stage.

Reproductive System of the Male Mouse and Examination of Spermatozoa

It is not often appreciated by students just how easy it is to obtain living mammalian spermatozoa for examination under the microscope. All that is required is a knowledge of the anatomy of the male reproductive tract and some very simple apparatus. Unlike oogenesis in the female, spermatogenesis in the male is a continuous process and sperm can normally be obtained at any time from sexually mature laboratory rodents such as the rat and mouse. In other male mammals, particularly wild ones, seasonal, hormonally induced changes in the testis may affect sperm production. Spermatozoa commence their development as spermatogonia, which line the basal region of the wall of the seminiferous tubules in the testis. These cells divide mitotically and some of the daughter cells that are produced become transformed into primary spermatocytes by enlargement and movement towards the tubule lumen. Primary spermatocytes then undergo the first meiotic division to form secondary spermatocytes, and these in turn undergo the second meiotic division to form spermatids. Thus from one primary spermatocyte containing the diploid number of chromosomes (40 in the mouse) arise four spermatids containing the haploid number (20). The spermatids then undergo a remarkable differentiation process that transforms them into cells whose organization is adapted to reaching, penetrating and activating the egg, and to transmitting the paternal genes. It is with these functions in mind that the mature spermatozoon should be viewed. All stages in the development of spermatozoa can be seen histologically in stained transverse sections of the testis and we recommend that students make themselves familiar with such stained material before examining living spermatozoa. Details of the transformation of spermatids into free swimming spermatozoa, which are not easy to observe without recourse to special staining techniques, are well described in Balinsky (1965).

A description of the dissection of the reproductive system

of the male mouse is included here, since we feel that attention should be drawn to the fact that because fertilization in mammals is internal associated organs are needed both in order to produce secretions (the seminal fluid) essential for the survival and motility of sperm, and for their deposition within the female reproductive tract.

Male reproductive system (Fig. 8.5)
Mature male mice are killed by means of chloroform (see p. 196) and pinned out in a suitable dissecting dish. The penis should first be located and the prepuce pushed back with forceps so as to expose the glans penis. Protruding from the external urethral orifice will be seen a thin cartilaginous rod. This is called the baculum or os penis, and serves to support the penis during intromission. It is present in many mammals, particularly carnivores. The glans penis has a rough appearance due to the presence of low epidermal papillae. The skin overlying the penis should be carefully removed. This will expose a pair of large, flat and leaf-shaped preputial glands just proximal and closely adherent to the glans penis. Their long secretory ducts open into the preputial cavity near the extremity of the prepuce and secretions from the preputial glands are presumed to contribute to the seminal fluid at ejaculation.

The skin should now be removed from the body wall and pinned back. In some cases the testes may be observed at this stage of the dissection, lying in an inguinal position. This is their normal position (there are no true scrotal sacs) but often they become retracted into the abdominal cavity during the killing procedure. A midline cut should now be made in the body wall extending from just above the penis to the base of the rib cage. This is continued on either side around the rib cage and the flaps of body wall pinned back. Lying above the bladder will be observed the very conspicuous seminal vesicles. Their function is not, as the name implies, to store semen, but to provide accessory secretions, in particular those contributing to the formation of the plug at copulation. Proteins

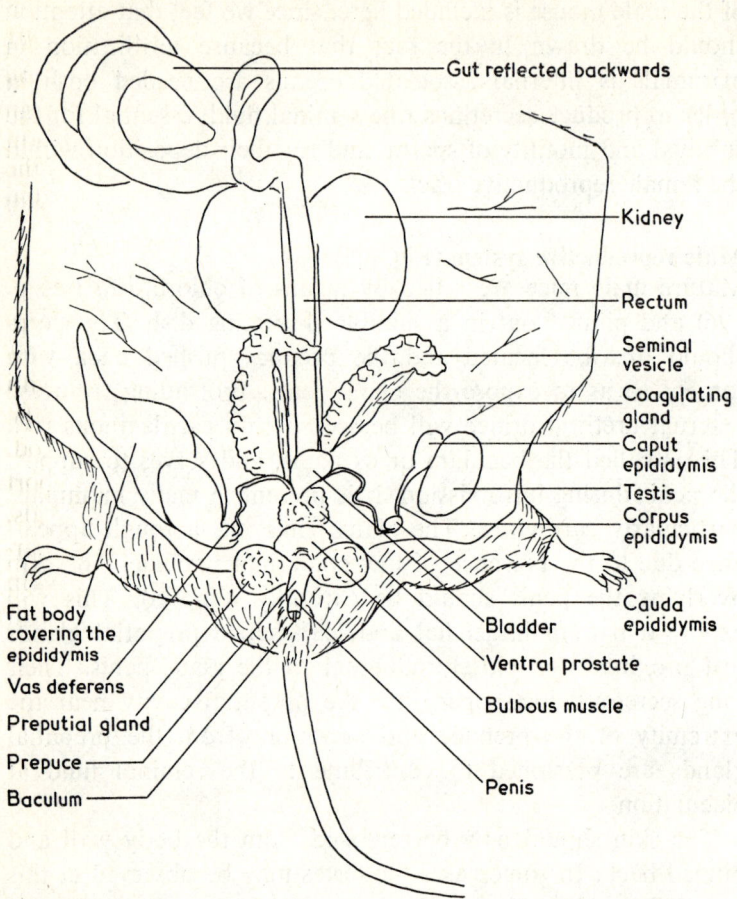

Fig. 8.5. *Drawing of the partially dissected male reproductive system.* The fat body has been removed from around the epididymis of the left testis. The uretha has not yet been traced through to the penis, or the bulbo-urethral glands isolated.

present in the seminal vesicle secretions are acted upon by a coagulating enzyme (called vesiculase) secreted from the coagulating gland. The latter is a paired structure lying attached to the inner margin of the seminal vesicles, and is part of the prostate gland. Two other pairs of diffuse glands, the dorsal and ventral prostates, lie above and below the urethra. Surrounding the urethra is a stout bulbo-urethral muscle. Contractions of this muscle lead to ejaculation.

In order to trace the urethra through to the penis it is necessary to cut through, and pare away, the pelvic girdle. Lying deeper down and outside the body of the bulbous are a pair of bulbo-urethral glands (Cowpers glands). Their ducts run through the muscle into the urethra in the region of the bulbous. The penis will be observed to consist of two separate bodies called the corpora cavernosa.

Besides providing a vehicle for the transport of sperm the secretions of the accessory reproductive organs (comprising the ampulla, seminal vesicles, coagulating glands, dorsal and ventral prostate, bulbo-urethral glands and preputial glands) contain metabolites such as inorganic ions, sugars and fatty acids etc., which are important for the motility of sperm; indeed, sperm only become motile when they leave the efferent ductules. The accessory gland secretions are also an important source of pharmacologically active substances (e.g., prostaglandins) which, by causing contractions of the uterus, help in mechanically propelling sperm into the oviducts.

The testes should now be located and if they lie in the inguinal canal, they should be drawn through onto the floor of the body wall. They are surrounded by a connective tissue sheath called the tunica albuginea. Within the testis the seminiferous tubules become collected into a number of excretory ducts called the rete testis. Efferent ducts (the vasa efferentia) emanate from the rete testis and unite externally to form a single duct called the ductus epididymis. This is divided roughly into three parts – the caput (head), corpus (body), and cauda (tail). A large fat body surrounds the caput edpididymis and vasa efferentia and it must be removed in order to display these structures. The ductus epididymis can now be traced down to the cauda epididymis, which is an elaborately coiled structure. As the ductus epididymis leaves the cauda it becomes the vas deferens, and is readily visible from its white appearance. Associated with the vas deferens are the spermatic artery and vein. The vas deferens should be traced to its terminal dilation, the ampulla. This is glandular in nature and narrows before entering into the posterior wall of the urethra close to the neck of the bladder.

Examination of spermatozoa
Spermatozoa can conveniently be obtained from the epididymis and/or vas deferens for examination under the microscope. Within the epididymis spermatozoa undergo a process known as 'sperm ripening' and this is manifested in the movement of a structure called 'cytoplasmic droplet' from the anterior to the posterior end of the mid piece, and eventually in its disappearance. The cytoplasmic droplet represents all that remains of the Golgi apparatus and other cytoplasmic constituents when spermateleosis (transformation of the spermatid into a spermatozoon) is nearing completion. Its movement and eventual disappearance is the culmination of shrinkage and dehydration processes that sperm undergo during maturation within the epididymis. By examining sperm from the caput, cauda, and vas deferens, differences in position and/or presence of the cytoplasmic droplet can be seen (Fig. 8.6). In order to obtain sperm, these structures are simply removed (a short segment in the case of the vas deferens) and chopped up with fine scissors in 0.85 per cent saline contained in a solid watch glass. With the aid of a Pasteur pipette a small drop of sperm suspension is then placed on a clean microscope slide and a coverslip applied. It is best to view the sperm under a high power phase contrast microscope, but considerable detail can also be observed by means of ordinary objectives if careful adjustment of the microscope iris diaphragm is made so as to cut down light and increase contrast.

Each spermatozoon consists of a head, mid piece and tail and has an average length of about 130 μm. The head of the mouse and rat spermatozoon has a very characteristic hook at its anterior end due to the shape of the acrosome which lies over the nucleus. The acrosome functions as a source of lytic enzymes (in particular, hyaluronidase in mammals) which aid the sperm in penetrating through the cumulus oophorous that

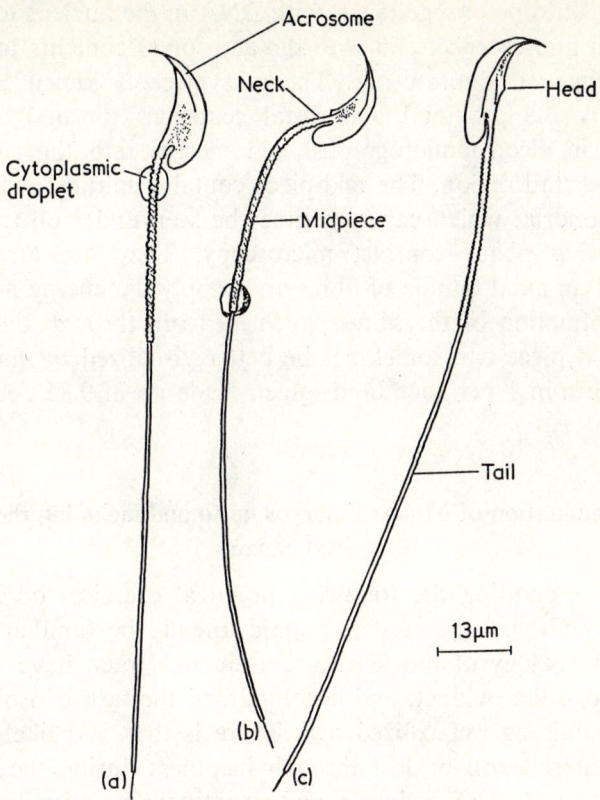

Fig. 8.6. *Drawings of typical mouse sperm obtained from* (a) the caput epididymis; (b) the cauda epididymis; (c) the vas deferens. Viewed under phase contrast, oil immersion objective.

surrounds the ovum (Plate 8.2a). Lytic enzymes also aid penetration through the zona pellucida and vitelline membrane. The acrosome is derived during spermateleosis from part of the Golgi apparatus, an organelle well recognized for its role in packaging hydrolytic and proteolytic enzymes. The acrosome can very readily be distinguished from the nucleus in living spermatozoa by mixing a drop of sperm suspension with an equal volume of 1:20 000 acridine orange (made up in 0.85 per cent saline) on a microscope slide, applying a coverslip, and viewing under u.v. light in a fluorescence microscope. Acridine orange reacts with DNA in the nucleus to give a green fluorescence, and with the acrosomal contents to give an orange-red fluorescence. The mid piece is joined to the head by the proximal and distal centrioles (normally only visible in electronmicrographs) and merges into the slightly narrower tail region. The mid piece contains spirally arranged mitochondria, which can sometimes be seen under oil immersion using phase contrast microscopy. They are arranged around an axial bundle of fibres and supply the energy needed for contraction of the fibres which also run through the tail. The mid piece can sometimes be better visualized by staining the sperm in 1 per cent Janus green made up in 0.85 per cent saline.

Examination of Mouse Embryos up to and including the Blastocyst Stage

Before attempting the following practical exercises on living mouse embryos, the student should already be familiar with the morphology of the female reproductive tract, have learnt to remove the oviduct, and accomplished the task of isolating and examining unfertilized ova. There is then less likelihood that material will be lost through ineptness during the more interesting task of isolating and examining post-fertilization embryos. The events following fertilization of the mouse egg, up to the time of blastocyst formation, are briefly outlined below. Reference should be made to Austin (1961) for a more detailed account of the development of the mammalian egg.

Fertilization involves the entry of a single sperm through the zona pellucida and egg membrane, and causes the egg to become activated. A renewed activity in the maternal chromosomes is initiated, and they complete their second meiotic division. Sperm penetration also causes the vitellus to contract slightly and the zona pellucida to expand, so that when the second polar body is extruded (an event normally dependent upon fertilization) it comes to lie next to the first polar body between the zona pellucida and vitellus in what is now the perivitelline space. In the mouse, the first polar body usually degenerates after formation of the second polar body (see Plate 8.2b). Whilst a number of other sperm may sometimes gain access to the perivitelline space, there is no further penetration of the egg membrane. Changes have occurred which prevent this, and such changes constitute what is described as a 'block to polyspermy'. Within the vitellus the sperm head swells and becomes the male pronucleus. It contains prominent nucleoli. The female pronucleus, which is smaller and also contains nucleoli, arises by formation of a membrane around the interphase haploid maternal chromosomes. Male and female pronuclei move towards each other and fuse (Plate 8.2c) whereupon the haploid chromosomes condense out and intermingle, so restoring the diploid chromosome number.

Mitotic division now takes place and results in the formation of two blastomeres almost equal in size. To reach the two cell stage from the time that copulation occurs takes about 24 hours but the next division of the blastomeres occurs more quickly and by about 36 hours (depending upon the strain of mouse) the four cell stage may have been reached. Division is not always synchronous between blastomeres and three and seven cell stages are sometimes seen. Eight cell stages may be reached within 48 hours and during this time the dividing egg has passed along the oviduct, propelled by the action of ciliated cells that line the oviduct lumen. When the egg reaches the 16 cell stage it is called a morula, and consists of a ball of inner and outer lying cells. It now passes from the oviduct into the uterus.

Soon after entering the uterus, at about three to three and a half days *postcoitum,* the morula becomes transformed into a blastocyst. A fluid filled cavity forms between the inner lying cells of the morula and enlargement of this cavity causes a group of cells, which are destined to give rise to the embryo proper and its embryonic membranes, to be pushed to one pole. This group of cells is called the inner cell mass and the cavity, the blastocoele or presumptive yolk sac cavity. The blastocoele and the inner cell mass are enclosed by a single layer of cells known collectively as the trophectoderm. At the time of implantation the trophectoderm cells become transformed into invasive trophoblast cells. Mouse blastocysts remain free within the uterine lumen for about one day, during which time they become evenly spaced along the length of the uterus and sink down into crypts lying in an antimesometrial position (Plate 8.2j). When it is fertilized, the egg, including the zona pellucida, has a diameter of about 113 μm and undergoes little change in overall dimensions during the early stages of development until it implants, but the blastomeres become progressively smaller with each succeeding division until a limiting size is reached. Throughout these early developmental stages the embryo is enclosed within the zona pellucida and this is not shed until about four and a half to five days *postcoitum,* when implantation is due to commence. The zona pellucida probably functions so as to prevent blastomeres from being lost and eggs from fusing together. Unlike the eggs of other vertebrates, mammalian eggs become diffentiated into specialized areas, the trophectoderm and inner cell mass, before any of the primary germ layers have been formed i.e. ectoderm, endoderm and mesoderm. This is an adaptation to intra-uterine development, since the first requirement for the embryo is to establish an intimate apposition with the uterine epithelium in order to obtain nutritive material.

From what has been stated in this account, it will be appreciated that living mouse embryos at particular stages of development can be obtained simply by examining the appropriate part of the reproductive tract at the right time after mating has taken place. Because of the difficulty of obtaining large numbers of mice at different stages of pregnancy on one particular day, this practical exercise is suitable only for small groups of advanced students. Practical details are as follows.

Timing of matings
We find it desirable to have mice ranging from recently mated to four days *postcoitum,* available on the day that the practical is to take place. If this is to be on a Friday, a number

of mouse boxes are set up on the preceeding Sunday evening containing one male and four female mice. On the following Monday morning the female mice are examined for the presence of vaginal plugs. The plug is formed as a result of secretions from the male seminal vesicles and coagulating glands mixing in the ejaculate. When combined with the cornified epithelial cells present in the lumen of the vagina, it forms a solid structure. It fills the vagina from the cervix to the vulva and is best looked for by dilating the vulva with forceps. Plugs are shed from the vagina after about 18 to 24 hours. The day of finding the plug is recorded as day 0, and mice that have mated are placed together in one box. Matings are similarly set up on the following Monday, Tuesday, Wednesday and Thursday evenings, so that those mice with plugs present on Friday, Thursday, Wednesday, Tuesday and Monday morning will probably have the following stages of embryos respectively: recently fertilized eggs in the ampulla; two cell stages in the oviduct; eight cell stages in the oviduct; morulae in the oviduct or early blastocysts in the uterus; fully formed blastocysts in the uterus.

Removal of eggs from the oviduct
The procedure for obtaining recently fertilized eggs from the oviduct is essentially the same as that described for obtaining non-fertilized eggs described on p. 198. When the mouse has been killed and the body wall opened, the uteri will be seen to be distended with male semen. As previously described, the oviducts are excised and transferred to culture fluid or saline contained in a solid watch glass. Before removing the eggs, a fine pipette can be inserted into the wall of the ampulla and some of the fluid it contains withdrawn. Examination of this fluid under the microscope will show that it contains spermatozoa. Depending upon how long the eggs have remained in the oviduct, some dispersion of the cumulus cells may have occurred spontaneously, but further mechanical and/or enzymatic removal of cumulus cells is usually required. When cleared of cumulus cells, the eggs are mounted as pre-

viously described beneath vaseline coated coverslips. Fertilized eggs are identified as such by the presence of two polar bodies, (or one intact and one degenerate polar body) lying in the perivitelline space (Plate 8.2b) but some rotation of the egg may be required in order to make them visible. When mounted, eggs can be rotated by carefully moving the coverslip with a needle, or, when in culture fluid under the dissecting microscope, simply by moving them with a fine seeker. High power phase contrast microscopy will frequently reveal the male and female pronuclei (Plate 8.2c) and occasionally suitable preparations are found in which the sperm tail is visible within the vitellus.

Oviducts removed from mice at one, two and three days *postcoitum* should be cut up into short pieces with fine scissors after they have been placed in culture fluid. This in itself is usually sufficient to liberate most of the contained eggs. In order to see the embryos amongst the cell debris it is necessary to use transmitted light and to adjust the reflecting mirror of the dissecting microscope so that the best contrast is achieved. If embryos cannot be found after initial segmentation of the oviducts, individual segments should be stroked with a blunt needle so as to squeeze out their contents. Eggs should be removed from amongst the debris by means of a mouth controlled suction pipette and transferred to fresh medium contained in a solid watch glass so that more detailed observations can be made.

Removal of blastocysts from the uterus
This is achieved by flushing culture fluid through the excised uterine horns. The technique should be practised first on the uteri of mice at one and two days *postcoitum*. Each uterine horn is removed by cutting across it just above the junction with the uterine corpus and by cutting along the mesometrium. It is then placed on a piece of filter paper and the remaining mesometrium trimmed off as close to the uterine wall as possible. This is in order to reduce contamination with blood which will obscure the blastocysts when they have been flushed

MAMMALS

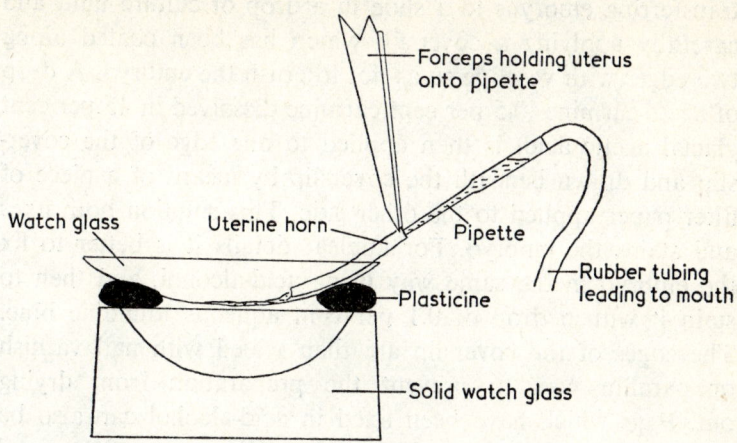

Fig. 8.7. Diagram showing method of flushing morulae and blastocysts from the isolated uterine horn.

out. A short, blunt-ended pipette, connected to the mouth by narrow rubber tubing, and filled with culture medium is inserted into one end of the uterus that has been cleanly cut. Holding the uterus onto the pipette with forceps, it is then transferred to the side of a watch glass mounted on a solid watch glass with plasticine (Fig. 8.7). Fluid is then forced through and out of the other end of the uterus by blowing, so that it accumulates in the centre of the watch glass. Care should be taken to avoid letting the uterus itself come into contact with the expressed fluid. The watch glass, mounted on the solid watch glass for support, is then transferred to the platform of the dissecting microscope and the flushings examined for the presence of blastocysts or morulae. The microscope should be focussed on the bottom of the watch glass and not at other levels, since the blastocysts soon sink down through the culture fluid. They have the appearance shown in Plate 8.2g, 8.2h and 8.2i.

Fixing and staining embryos

Nuclear and cytoplasmic details are often better observed after the embryos have been fixed and stained. This is achieved by transferring embryos to a slide in a drop of culture fluid and carefully applying a coverslip which has been coated along two edges with vaseline so as not to crush the embryo. A drop of aceto-carmine (0.5 per cent carmine dissolved in 45 per cent glacial acetic acid) is then applied to one edge of the coverslip and drawn beneath the coverslip by means of a piece of filter paper applied to the other side. This solution both fixes and stains the embryo. For nuclear details it is better to fix the embryo in the same way using acid-alcohol, and then to stain it with a drop of 0.1 per cent aqueous toluidine blue. The edges of the coverslip are then sealed with nail varnish or paraffin wax to prevent the preparation from drying out. Eggs which have been fixed in acid-alcohol can also be stained with a 1:20 000 solution of acridine orange in normal saline, and then examined by fluorescence microscopy. In fixed material acridine orange combines with DNA to give a bright green fluorescence, and with RNA to give a red fluorescence, when stimulated with ultra-violet light. Thus the nucleus and nucleolus can be readily picked out. The cytoplasm gives a yellowish-red appearance due to the presence of ribosomal material.

Examination of stained sections

Although the emphasis here has been on living embryos, much information about the development of the mammalian embryo and especially its relationship to the reproductive tract, can be gained by examination of serial stained sections. These are obtained simply by removing the appropriate part of the reproductive tract (oviduct or uterus) at the required time after mating, and then fixing, embedding and sectioning the tissue in the usual way.

In Vitro Culture of Mouse Eggs and Egg Fusion Technique

The study of early mammalian development has expanded rapidly in the last decade, largely due to the introduction of techniques which enable the mammalian egg to be cultured and manipulated *in vitro*. It is now possible to fertilize mouse eggs outside the body of the mouse and also to grow fertilized eggs to the blastocyst stage in a chemically defined medium. Such blastocysts can then be transferred to the uterus of a recipient female mouse that has undergone a sterile mating. Within such a receptive uterus the blastocyst will implant, and in at least a proportion of cases, give rise to perfectly normal offspring which are themselves capable of reproducing. This shows that culturing eggs *in vitro* need not adversely affect their subsequent development *in utero*. The techniques used in these procedures are described in great detail by experts in the field, in a book edited by Daniel (1971). Human eggs have also been grown to the blastocyst stage after fertilization *in vitro*, (Steptoe, Edwards and Purdy, 1971), although no successful implantations have so far been reported after intra-uterine transfer.

In vitro culture of mammalian eggs has also led to the development of new ideas concerning the mechanism controlling trophoblast and inner cell mass differentiation in the blastocyst. Earlier workers maintained that the uncleaved mammalian egg had a pre-existing heterogeneity which could be visualized by differential cytochemical staining of 'ventral' and 'dorsal' areas (Dalcq, 1957; Mulnard, 1961). Such heterogeneity was thought to reflect the presence of morphogenetic factors which, during the first two cleavage stages, became segregated into different blastomeres and therefore predetermined which would differentiate into trophoblast and which into inner cell mass. A number of elegant experiments have now been performed which suggest that this might not be true. From the results of initial experiments carried out by Tarkowski and Wroblewska (1967), in which the fate of each blastomere of four and eight cell stage mouse eggs was followed by dissociating the egg and growing each blastomere separately in culture, it was proposed that epigenetic factors, such as cell position, rather than pre-determined intrinsic factors, might govern which blastomeres differentiated into trophoblast and which into inner cell mass. Confirmation of this idea has come from experiments in which: (a) mouse eggs labelled with H^3 thymidine have been fused to unlabelled eggs so as to completely enclose what would normally have been trophoblast forming blastomeres, and the fate of such blastomeres determined by autoradiography after culture of the giant composite embryo to the blastocyst stage (Hillman, Sherman, and Graham, 1972); (b) blastomeres have been labelled with vital markers and their fate followed cinematographically during culture to the blastocyst stage (Wilson, Bolton and Cuttler, 1972; Stern and Wilson, 1972); (c) mouse blastocysts and eight cell stages have been disaggregated so as to disrupt spatial relationships and the blastomeres allowed to reform in culture to form blastocysts of mixed composition (Stern, 1972). All these experiments show that the blastomeres are considerably labile, even in the early blastocyst stage.

Also of considerable interest has been the production of genetically mosaic mice (so-called chimaeras, or allophenic mice) by fusing together eggs derived from parents of different strains, growing the embryos to the blastocyst stage, and transferring them to the receptive uterus of a female mouse (Mintz, 1965; 1971a). Chimaeras have also been produced by injecting inner cell mass cells, derived from the blastocyst of one mouse strain, into the blastocoelic cavity of another mouse strain blastocyst (Gardner, 1968). By chance, potential male and female mouse eggs have been fused together and after culture and transfer to a receptive uterus, shown to give rise to offspring that are hermaphrodites (Tarkowski, 1961; Mystakowska and Tarkowski, 1968). The study of chimaeras is now opening up a new dimension in mammalian development, enabling the interaction of cells of different genotype to be followed and the clonal basis of development to be investigated, in a way which has hitherto been impossible. It is also bringing the possibility of genetic engineering a little closer to reality.

In keeping with our view that a modern practical course in animal development should, wherever possible, reflect current advances in knowledge, we feel that students ought to have experience of the methods used for culturing and manipulating mammalian eggs. The culture of eight cell stage mouse eggs to blastocysts, removal of the zona pellucida, and subsequent fusion of eggs, in our experience constitutes a suitable practical exercise for advanced students.

Culturing eight cell and earlier stage eggs
Although it is possible to culture mouse eggs from postfertilization up to the blastocyst stage in a simple chemically defined medium (Whitten and Biggers, 1968), the chances of success under teaching laboratory conditions are greater if older mouse embryos are used at the commencement of culture. It is also an advantage to use eggs obtained from hybrid matings since the acquired hybrid vigour helps the egg to withstand the traumas of *in vitro* culture. If there is a likelihood that eggs earlier than eight cell stages will be obtained and attempts made to culture them, then it is best to use a medium containing pyruvate and lactate, which are the energy sources required for division by at least the two cell stage and the undivided egg. The medium devised by Whitten (1971) is a suitable one to use and is prepared as follows:

Component	*Concentration in mg*
NaCl	514
KCl	36
KH_2PO_4	16
$MgSO_4 7H_2O$	29
$NaHCO_3$	190
Na Pyruvate	3.5
Ca Lactate $5H_2O$	83
Glucose	100
K Penicillin G	8
Streptomycin	5

These ingredients are placed in a suitable bottle and 100 ml of de-ionized glass distilled water are added. After solution has occurred, 0.37 ml of 60 per cent sodium lactate syrup and 0.1 ml of 1 per cent phenol red indicator are added, and then 300 mg of bovine serum albumin. A gas mixture consisting of

5 per cent CO_2, 5 per cent O_2 and 90 per cent N_2 is then bubbled through the medium for one hour and the mixture is then sterilized by filtration through a millipore filter (pore size 0.45 μm). The culture medium should have a pH of 7.2 to 7.4 and any change from this warrants further gassing. It is best distributed into sterile bijoux bottles (5 ml) using sterile syringes and needles and further gassing of the medium should be carried out before use. This is done by making two holes in the screw top of the bottles, one to admit a needle connected to the gas cylinder tubing, and the other to admit a needle open to the atmosphere which allows gas to escape after it has bubbled through the medium, so avoiding frothing.

For culture of eight cell stage and older embryos, some workers prefer to use Earle's balanced salt solution (Earle's BSS), supplemented with serum. This has the advantage, especially from the point of view of practical class work, that it can be bought ready made and sterile (Wellcome Laboratories). Heat inactivated calf serum (Wellcome Laboratories) is added to the Earle's BSS with a sterile syringe and needle so as to give a 30 per cent solution. The medium is then gassed and distributed into smaller bottles as described above. It is warmed to 37°C before use.

Dissecting instruments, solid watch glasses and micropipettes, should be sterilized by wrapping them in aluminium foil and heating to 150°C in a sterilizing oven. The working area should be made as clean as possible. Eggs are removed from the oviduct as described on p. 210 but instead of using chloroform mice are killed by dislocating the cervical vertebrae (Scott and Ray, 1972) and the body surface cleaned with cotton wool soaked in alcohol. The oviducts will contain eggs at the eight cell stage (see Plate 8.1e) at about two and a half days *postcoitum*. Eggs should be transferred as quickly as possible from the medium in which the oviducts have been cut up, to fresh medium with as little transfer of cell debris as possible. Eggs often fail to develop further because they are left too long on the stage of the dissecting microscope under unsuitable conditions. They should then be transferred from fresh

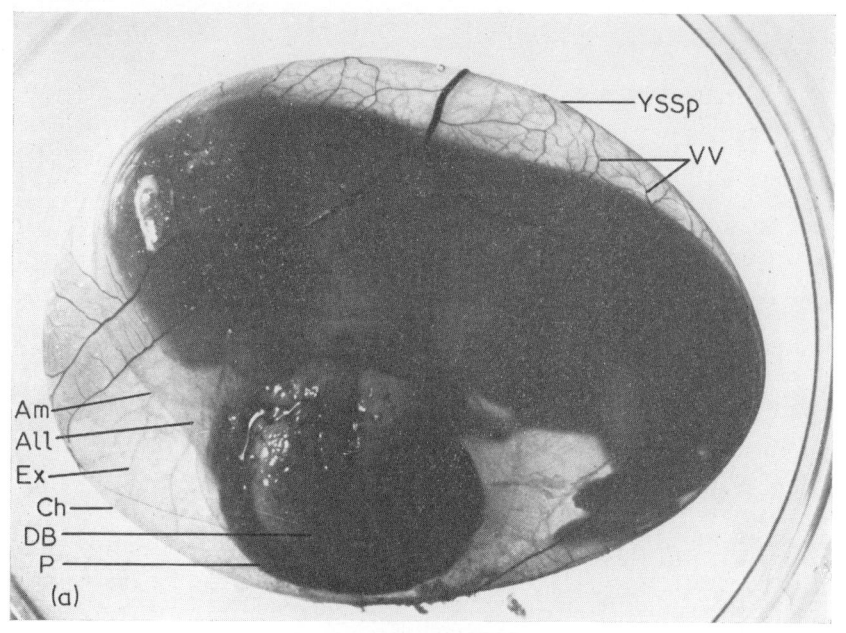

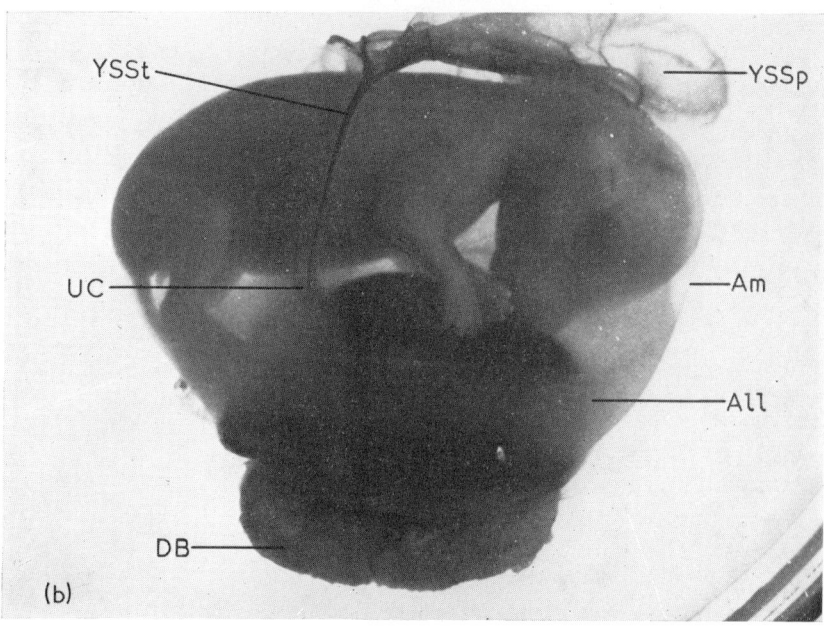

Plate 8.1 Photographs of a rabbit conceptus to show (a) its appearance when removed intact from the pregnant uterus and (b) its appearance when the yolk sac splanchnopleur has been cut away. YSSp – Yolk Sac Splanchnopleur; VV – Vitelline Vessels; Ch – Chorion; P – Chorio-Allantoic Placenta; DB – Decidua Basalis; Am – Amnion; Ex – Exocoel; All – Allantois; YSSt – Yolk Sac Stalk; UC – Umbilical Cord.

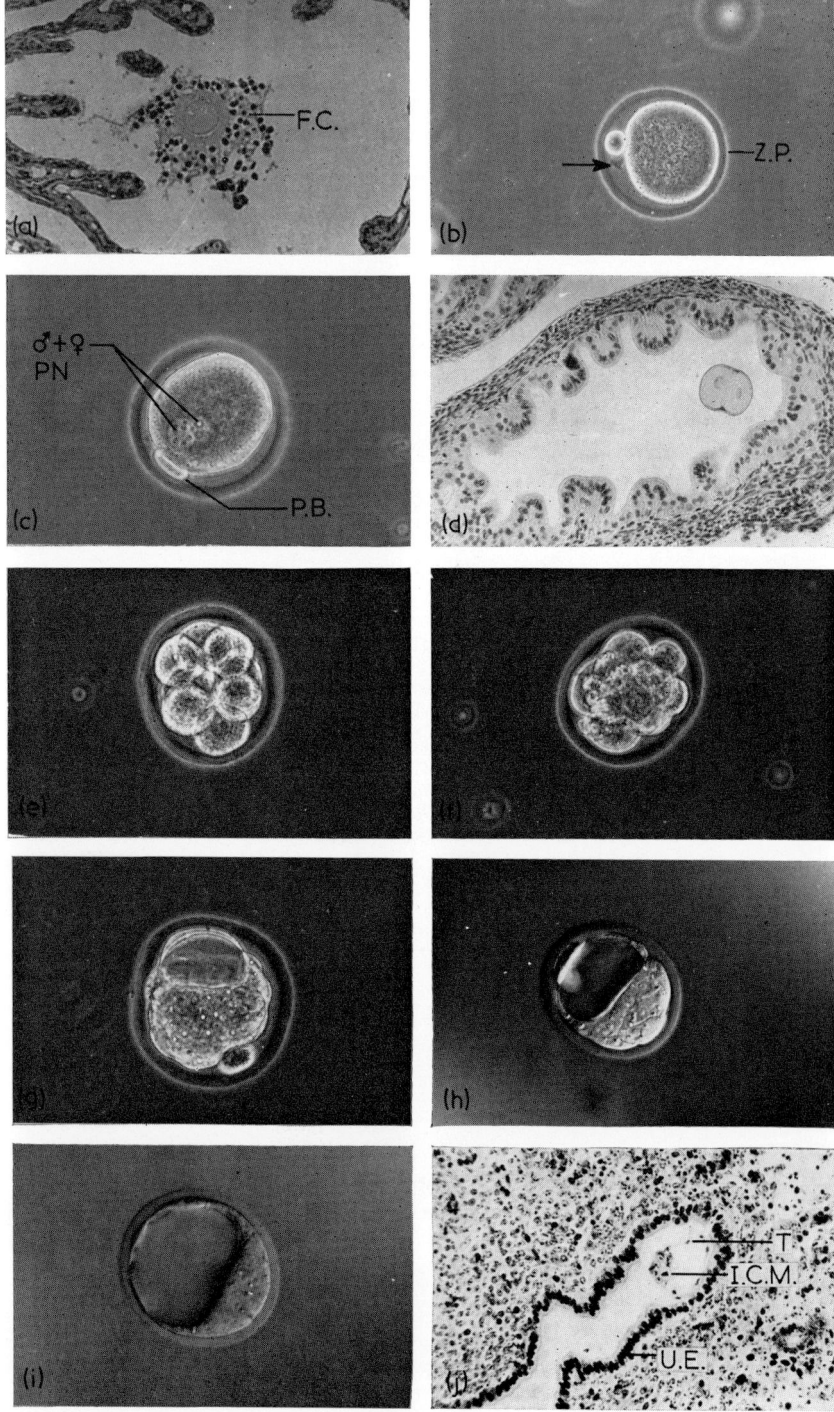

Plate 8.2 Stages in the early development of the mouse embryo

(a) Transverse section through the ampullary region of an oestrous mouse oviduct. An unfertilized ovum surrounded by follicle cells (FC) lies free within the lumen of the ampulla. Stained with haematoxylin and eosin. ($\times$ 100)

(b) Living fertilized ovum isolated from the ampulla at about 10 h after mating. The first polar body has disintegrated (arrowed) and the second polar body is visible in the pervitelline space lying between the zona pellucida (ZP) and the vitellus. ($\times$ 200 phase contrast).

(c) Living fertilized ovum at about the same stage as (b) but compressed more in order to reveal the male and female pronucleus ($\male$ + $\female$ PN) lying close together. The second polar body (PB) is also visible. ($\times$ 200 phase contrast).

(d) Transverse section through the oviduct of a mouse at about 24 h after mating. A two cell stage embryo is visible in the lumen of the oviduct. Note that the lumen is much narrower than it is in the ampullary region. The zona pellucida appears very much less distinct than it does in the living embryo. This is due to shrinkage during fixation and subsequent histological processing. Stained with haematoxylin and eosin. ($\times$ 100).

(e) Living eight cell stage embryo removed from the oviduct of a mouse at about $2\frac{1}{2}$ days after mating. ($\times$ 200 phase contrast).

(f) Living morula obtained by flushing the uterus of a mouse at about 3 days after mating. ($\times$ 200 phase contrast).

(g), (h) and (i) Living blastocysts, early, mid, and late, respectively. These were obtained by flushing the uteri at about $3\frac{1}{2}$ to 4 days after mating. Note the progressive enlargement of the blastocoele and the localization of the inner cell mass at one pole. ($\times$ 200 phase contrast).

(j) Detail of a transverse section through the uterus of a mouse at about $4\frac{1}{2}$ days after mating, showing a blastocyst lying within the uterine lumen at an implantation site. The zona pellucida has been shed but the trophoblast cells (T) have not yet penetrated between the uterine epithelial cells (UE). The inner cell mass (ICM) lies opposite to the point where the trophoblast cells closely adhere to the uterine epithelium. This is the normal position for the embryonic pole to occupy at the time of implantation. Stained with Heidenheim's haematoxylin. ($\times$ 100).

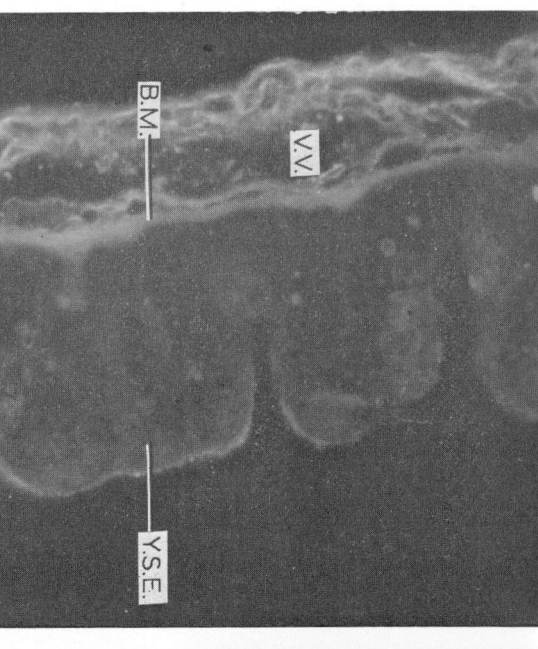

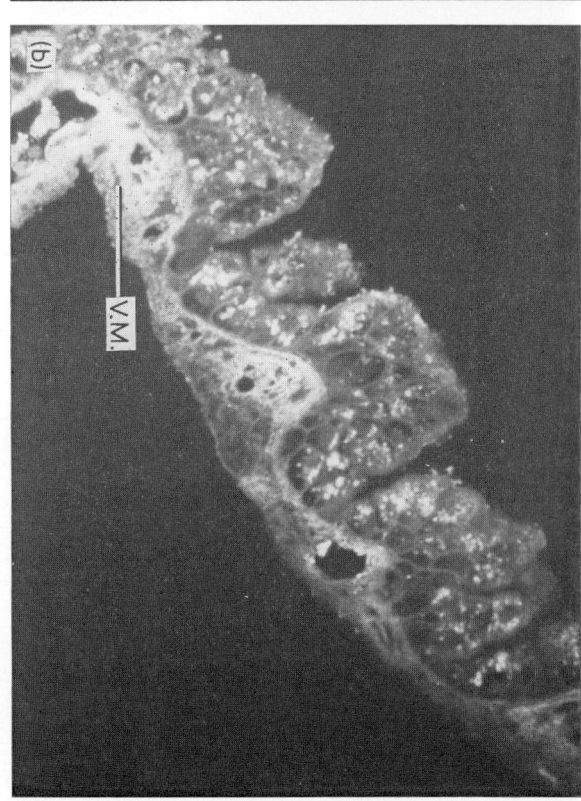

Plate 8.3 (a) Detail of a transverse section of rabbit yolk sac splanchnopleur that had been treated with FITC labelled sheep anti-rabbit immunoglobulins. Specific fluorescence indicative of rabbit immunoglobulins is weak in endocytosis vesicles present in the yolk sac endoderm (YSE) but very strong in the basement membrane (BM). Much of the fluorescence in the vitelline vessels (VV) is due to autofluorescing erythrocytes. Photographed using a dark field condenser and Tri–X film with a 1½ min exposure. (× 300). (b) A different area of rabbit yolk sac splanchnopleur treated in the same way as (a). Here the yolk sac endoderm contains many intensely fluorescing vesicles indicative of rabbit immunoglobulins that have been taken up from the uterine fluid. There is also strong specific fluorescence in the vascular mesenchyme (VM) probably indicative of immunoglobulins that have in some way been transported intact across the endodermal cells and basement membrane. Photographed as in (a). (× 200).

medium to the culture vessel (Fig. 8.8). This consists of a sterile plastic disposable petri dish (18×50 mm) half filled with liquid paraffin that has been treated as follows. Several days before the class is to commence, liquid paraffin is sterilized using dry heat and 400 ml mixed with 20 ml of culture fluid. The gas mixture is then bubbled through it, taking care to maintain

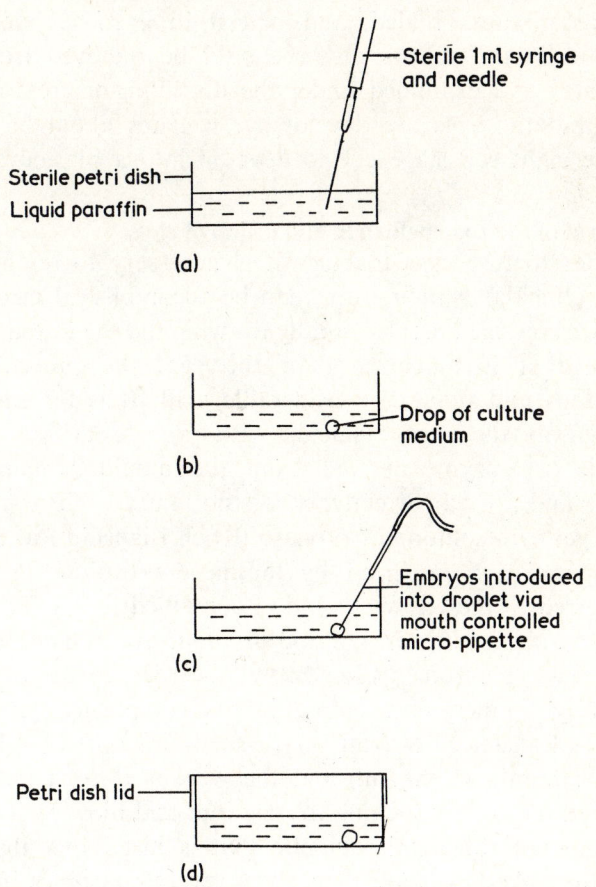

Fig. 8.8. *The culture vessel and method of transferring eggs to it.* In (a) and (b) the method of transferring a small volume of culture fluid beneath the liquid paraffin is shown. Embryos are then transferred to this drop from a mouth controlled micro-pipette as in (c). The culture vessel complete with lid (d) is then ready for transfer to a dessicating jar.

sterile conditions, and the equilibrated liquid paraffin is stored under gas in a suitable sealed container. Four drops of culture fluid (about 0.1 ml) are delivered from a mouth controlled micropipette under the liquid paraffin. Working under the dissecting microscope, eggs are then transferred to each of these drops. The culture vessels are then placed in an empty vacuum dessicator (without dessicant) which is filled with the gas mixture, sealed, and placed in a 37°C incubator. Periodically the culture dishes should be removed from the dessicator and examined under the dissecting microscope for signs that the eggs are developing. It takes about 36 hours for the eight cell stage egg to develop into a blastocyst.

Removal of the zona pellucida and fusion of eggs
In order to fuse eggs together it is necessary to remove the zona pellucida. Although this can be accomplished mechanically (Tarkowski, 1961) by rapidly drawing the egg into a micropipette that is narrower than the egg, the procedure is hazardous and requires considerable skill. It is far easier to digest away the zona pellucida with pronase (Mintz, 1962). This is an enzyme derived from the mould *Streptomyces gracilis* and if used correctly, causes no injury to the egg itself. A 0.5 per cent solution of pronase (Koch Light) in Earles BSS is prepared and sterilized by millipore filtration. Eggs are transferred to the enzyme solution contained in a solid watch glass and observed over a period of about five minutes, during which time the zona pellucida will be seen to become progressively thinner. Just before it has completely lyzed, the eggs are transferred to fresh sterile medium containing bovine serum albumin (1 mg/ml) and the weakened zona pellucida detached by rapidly sucking the egg into and blowing it out of a mouth controlled micropipette that is just larger than the egg. The naked eggs are then given several washings in fresh culture medium in order to remove any traces of pronase, and then transferred in pairs to drops of serum containing medium under liquid paraffin. Naked eight cell eggs can often be made to adhere at room temperature simply by bringing them into

close contact with fine needles. With earlier stages, however, it is necessary to warm up the eggs to 37°C. This can be done quite effectively by placing the culture vessel on a temperature controlled hot plate. In order to keep the naked eggs in close contact, the culture medium is withdrawn by means of a micropipette until the eggs are forced together. After warming the culture chamber at 37°C, the culture medium is then added back to the droplet, taking care not to introduce air bubbles, and the composite embryo checked to ensure that the blastomeres have not fallen apart. Culture of the chimaera is then carried out as previously described. It must be appreciated that the technique described here is somewhat crude when compared to that used by experts in this field, and that to achieve development of maximum numbers of eggs, elaborate precautions outside the scope of teaching laboratory conditions, need to be taken. These are described in great detail by Mintz (1971b).

Arrangement of the Foetal Membranes in the Rabbit and Localization of Immunoglobulin in the Yolk Sac Splanchnopleur by means of Immunofluorescence

In order for growth and differentiation of the mammalian embryo to take place, metabolites have to be transferred to it from the mother. This is accomplished through specialized connections to which the term 'placenta' is given. Originally this referred to the flat cake-like structure which constitutes the chorio-allantoic placenta of rodents and man, but it has now come to have a wider meaning. Mossman (1937) has defined the mammalian placenta as 'an apposition or fusion of the foetal membranes to the uterine mucosa for the purpose of exchange'. This definition encompasses, in addition to the chorio-allantoic placenta, foetal membranes such as the chorion and yolk sac splanchnopleur. The yolk sac splanchnopleur in particular is a prominent organ of exchange in mammals such as the rabbit, rat, mouse and guinea pig. The manner in which the foetal membranes develop in the rabbit is depicted diagrammatically in Figs. 8.9, 8.10, 8.11, 8.12.

The rabbit blastocyst, unlike that of the mouse, spends a considerable period of time (about four days) free within the uterine lumen prior to implantation. During this time it accumulates fluid and eventually swells to a diameter of about 1 cm, with the result that the uterus becomes distended. Implantation occurs at about seven days *postcoitum* and is described as being 'central'. Over the surface of the blastocyst, knobs of invasive trophoblast are formed and these penetrate into the uterine epithelium (Fig. 8.9). Endodermal cells are

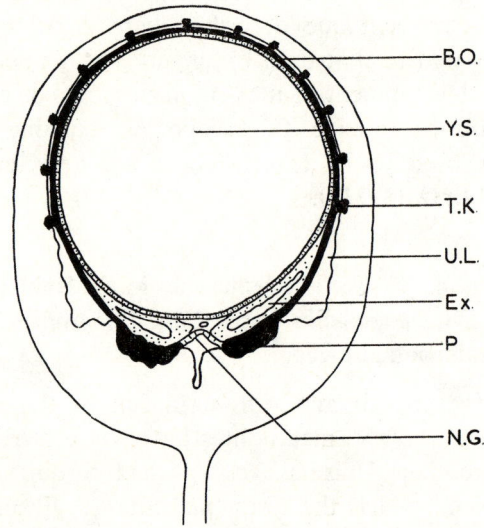

Fig. 8.9. *Diagram of a section through a rabbit conceptus at 8 days of gestation.* BO – Bilaminar Omphalopleure; YS – Yolk Sac Cavity; TK – Trophoblastic Knob; UL – Uterine Lumen; Ex – Exocoel; P – Area in which the chorio-allantoic placenta will form from trophoblast tissue which has extensively invaded the uterine mucosa; NG – Neural Groove.

delaminated from the inner cell mass (now transformed into the embryonic plate) and come to line the blastocoelic cavity, so converting it into the yolk sac cavity. The wall of the yolk sac at this stage is composed of outer trophoblast and inner endodermal cells and is referred to as the bilaminar omphalopleur. Gastrulation takes place in the embryonic plate in a manner similar to that occurring in the chick blastoderm (Balinsky, 1965) and results in mesoderm formation below the embryonic plate. Attachment and invasion of trophoblast tissue is extensive in those areas where the chorio-allantoic placenta will form, but in other areas connections are only temporary. Cavitation within the mesoderm results in formation of the exocoel (Fig. 8.10) and this extends between the

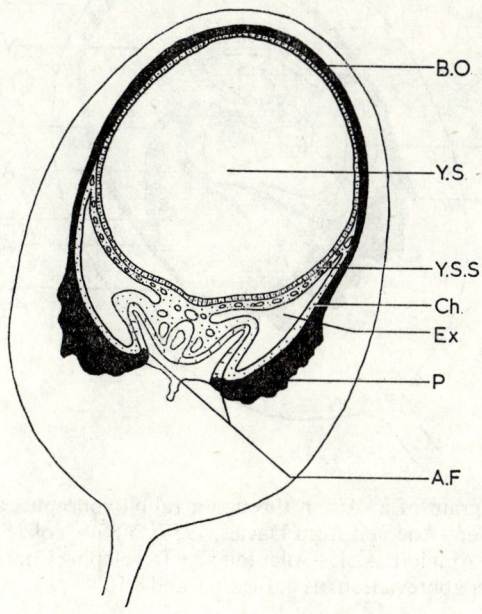

Fig. 8.10. Diagram of a section through a rabbit conceptus at about 9 days of gestation. (Adapted from Davies, 1958). AF – Amnion Folds. Ch – Chorion. Other abbreviations as in Fig. 8.9.

trophoblast and endoderm for part of the way, giving rise to the chorion (trophoblast and mesoderm) and to the amnion folds (embryonic ectoderm and mesoderm). The inner splanchnic wall of the yolk sac (endoderm and mesoderm) becomes richly vascularized and a vitelline circulation is established. This vascularized portion of the yolk sac wall is known as the yolk sac splanchnopleur. The chorion on the other hand, is largely non-vascular. With the fusion of the amnion folds the embryo gradually becomes enclosed in amniotic fluid and external to this, in exocoelomic fluid, and bulges into the yolk sac cavity. An outgrowth from the hind gut region of the embryo, the allantois, pushes its way across the exocoel towards the chorion (Fig. 8.11). The allantois in the rabbit is

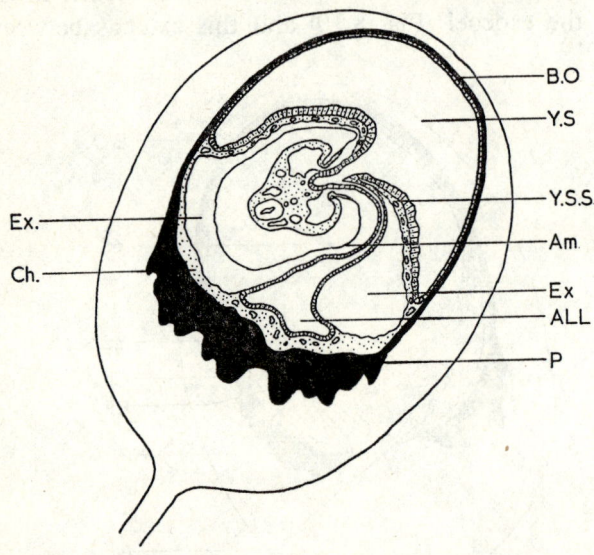

Fig. 8.11. Diagram of a section through a rabbit conceptus at about 12 days of gestation (Adapted from Davies, 1958). YSS – Yolk Sac Splanchnopleur; Am – Amnion; ALL – Allantois; P – Developing Chorio-Allantoic Placenta. Other abbreviations as in Figs. 8.9 and 8.10.

vesicular and well vascularized. It fuses with the chorion in the basal region, causing a further proliferation of trophoblast and allantoic mesenchyme. Maternal tissues and blood vessels are eroded away and the foetal blood vessels come into close apposition with the maternal blood, so forming the chorio-allantoic placenta. The endodermal cells of the yolk sac splanchnopleur are highly absorptive. Material present in the yolk sac fluid is endocytosed and the volume of fluid gradually decreases so that the inner and outer walls of the yolk sac come closer together. At about 13 days *postcoitum* the bilaminar omphalopleur breaks down and concomitantly a new uterine epithelium is established. What was formerly the yolk sac cavity now becomes the new uterine lumen, and the endodermal cells of the yolk sac splanchnopleur are exposed to the proteinaceous uterine fluid. The final arrangement of the foetal membranes are as depicted in Fig. 8.12.

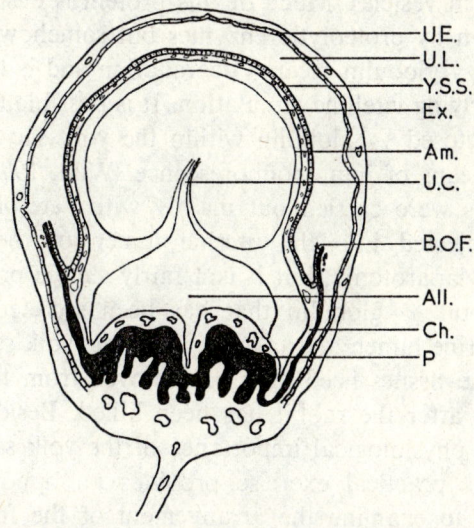

Fig. 8.12. Diagram of a section through a rabbit conceptus as it appears between 15 days of gestation and term. Details of the foetus are not shown. (Adapted from Davies, 1958). UE – Uterine Epithelium; UC – Umbilical Cord; BOF – Bilaminar Omphalopleur Fringe; P – Definitive Chorio-Allantoic Placenta. Other abbreviations as in Figs. 8.9, 8.10 and 8.11.

Besides breaking down endocytosed proteins to provide amino acids for the developing foetus, the rabbit yolk sac splanchnopleur, in the later stages of gestation, plays an important role in the transmission of passive immunity. (See reviews by Brambell, 1958; 1970, and by Wild, 1973b.) In normal circumstances foetal and newly born mammals do not actively produce their own antibodies, since they receive little in the way of foreign antigenic stimulation *in utero*. In order that the newborn mammal may have protection from invasive organisms when it enters the hostile world outside the uterus, antibodies, largely in the form of γ G – globulin, are transmitted from the mother. In some mammals, like the rabbit for instance, transmission takes place entirely before birth. Antibodies are secreted into the uterine fluid and taken up into the endodermal cells of the yolk sac splanchnopleur by pinocytosis. Together with other endocytosed proteins they accumulate in vesicles. Much of this protein is destined to be broken down by proteolytic enzymes but somehow a proportion of the γ-globulin escapes degradation and is transported to the underlying vitelline circulation. It is possible to visualize such endocytosed γ-globulin within the yolk sac splanchnopleur, by means of immunofluorescence (Wild, 1970; 1973a,). Such studies were carried out mainly with heterologous γ-globulins injected into the uterine lumen of the pregnant rabbit after laparotomy, but it is a fairly simple procedure to localize rabbit γ-globulin that has been secreted naturally into the uterine lumen and taken up into the yolk sac splanchnopleur. The tissues need only be removed from the isolated conceptuses after the rabbit has been killed. Besides demonstrating the physiological importance of the yolk sac splanchnopleur this practical exercise provides an opportunity for the student to examine the arrangement of the foetal membranes of the rabbit conceptus. Since pregnant rabbits are costly and provide a variable number of conceptuses, this exercise is suitable only for small groups of students. Adequate time should be available for tissues to be fixed, embedded, sectioned, and stained with fluorescent labelled antibodies.

Examination of the rabbit conceptus

Rabbits that are 24 to 26 days pregnant are required (gestation lasts for 30 days in the rabbit). Rabbits are coitus-induced ovulators. If mating occurs the doe will ovulate 10 to 12 hours later. A buck is placed in the cage of separately housed does and if the doe is receptive the buck is allowed to fall twice and then removed from the cage. The doe is killed at the appropriate time by injection of nembutal (sodium pentobarbitone – Abbot Laboratories) into the marginal ear vein at a dose of 1.5 ml per kg body weight. The fur over the abdominal region is removed either by means of electric clippers or by wetting it and cutting away the underlying skin. An apron made of lint or other suitable material, and containing a longitudinal slit, is laid over the abdomen. A mid-line incision is then made in the body wall and the uterine horns pulled out.

Starting at the utero-tubal junction, an incision is made in the wall of the uterus and carried forward in an antimesometrial position, care being taken to avoid large blood vessels or damage to the underlying foetal membranes. If the uterus is allowed to dry out at any time, the yolk sac splanchnopleur adheres too strongly to the uterine epithelium and consequently is easily torn. It frequently helps if, prior to dissecting out the conceptus (the foetus enclosed in its membranes), 3 to 4 ml of normal saline are injected into the uterine lumen at the utero-tubal junction and the fluid worked around the conceptuses by manipulation of the uterus. If the yolk sac splanchnopleur is free it will bulge through the incision due to the pressure of the contained foetal fluids. When all conceptuses of one horn have been exposed they are detached from the uterus by gently peeling away the decidua basalis (see Plate 8.1a). This is composed of maternal decidual tissue and has a whitish appearance. Each conceptus is placed in a petri dish containing normal saline and any contaminating blood and uterine debris is rinsed away. The yolk sac splanchnopleur must not be allowed to dry out. At this stage the intact conceptus has the appearance shown in Plate 8.1a.

The villous nature of the endodermal cells can easily be observed under a dissecting microscope, as well as the blood coursing through the underlying vitelline vessels. The paraplacental chorion is a largely avascular endoderm-free area, which separates the yolk sac splanchnopleur from the margin of the chorio-allantoic placenta. Just below the sinus terminalis (the circumferential terminal vitelline vessel) the paraplacental chorion is fringed by the remnants of the bilaminar omphalopleur. The vitelline circulation in the yolk sac splanchnopleur is connected to the main blood supply in the umbilicus via the yolk sac stalk. This consists of a main vitelline artery and vein running closely adherent to the amnion and through the allantois. It is best visualized by draining away the saline around the conceptus and cutting circumferentially around the paraplacental chorion with fine scissors. This will also expose the amnion and the allantois. The amniotic fluid often has a straw coloured appearance but if the foetus has defaecated may appear green due to the passage into it of biliverdin. Allantoic fluid on the other hand, often has a white, milky appearance, making it easy to see (Plate 8.1b). The yolk sac stalk should now be cut and the yolk sac splanchnopleur transferred to a fresh petri dish ready for fixation.

Preparation of tissue for treatment with fluorescent labelled antibodies
The yolk sac splanchnopleur is a delicate membrane and handling is made easier by stretching pieces of the membrane over rectangular windows cut in card. The card, with the membrane attached, can then be picked up with forceps. It is placed in 95 per cent ethanol that has been pre-cooled to 4°C and allowed to fix at this temperature overnight in a refrigerator. Dehydration is then carried out with three changes of precooled absolute ethanol (two hours in each change at 4°C). Clearing is then achieved by passing the membrane (still attached to the card) through three consecutive changes of pre-cooled xylene (two hours in each change, also at 4°C). Times are not critical and if necessary the tissue can be left overnight in xylene at 4°C. Wax (m.p. 56°C) is dissolved in the last change of xylene which is brought first to room temperature and then to 37°C. The membrane is then passed through four consecutive changes of wax (one hour in each), and before the final embedding, the membrane is carefully detached from the card. After the blocks have been trimmed, transverse sections are cut at 4 μm in the usual way on a microtome. Two or three sections are placed towards one end of a clean microscope slide appropriately marked with a diamond, and the sections flattened by floating them on water with the slide resting on a hot plate. Care should be taken to

ensure that the temperature does not exceed 40°C and the water on which the sections float should be drained away with filter paper as quickly as possible to avoid any leaching of antigen. The sections are then dried at 37°C for one hour. This will ensure adhesion of the sections during subsequent procedures providing adequate flattening has been achieved. Adhesives should be avoided since they may interfere with the fluorescent staining. The sections are de-waxed by passing them through three baths of pre-cooled xylene (one to two minutes in each bath). They are then passed through three baths of pre-cooled absolute ethanol to remove any trace of xylene and through three baths of pre-cooled 0.1 M phosphate buffered saline (PBS), pH 7.2 to remove ethanol. (PBS is prepared by dissolving 8.5 g of NaCl, 1.07 g of Na_2HPO_4 and 0.39 g of NaH_2PO_3 in distilled water, and making up to 1 l.).

Immunofluorescent staining
Localization of rabbit immunoglobulins is achieved by application of fluorescent labelled antibodies to the sections. The principle behind the fluorescent antibody technique is relatively simple and depends upon the biological specificity of antibodies (see Nairn, 1969 for a comprehensive review). Two approaches are possible:

In the first, fluorescein isothiocyanate (FITC) labelled sheep antibodies to rabbit immunoglobulins can be applied directly to the sections. Such labelled antiserum is best obtained commercially (Wellcome Research Laboratories) and is supplied as a freeze-dried powder (details of methods used to label antisera are well described in Nairn, 1969). The FITC labelled antiserum is reconstituted with distilled water and further diluted 1:5 with PBS. The slides are removed from the final bath of PBS and wiped free of liquid save for a small amount around the sections. A small volume of labeled antiserum is then applied to the buffer solution over the sections, either with a micropipette, or with a platinum wire loop. Controls are set up using either FITC labeled normal sheep γ – globulin fraction or unlabelled sheep anti-rabbit immunoglobulin (also obtainable from Wellcome Research Laboratories). The latter antiserum acts as a blocking agent (see below). The slides are now incubated on a level surface inside a humidity chamber for half an hour (sandwich boxes half filled with water and containing inverted glass staining troughs on which to rest the slides, make suitable humidity chambers). At no stage should the antiserum covering the sections be allowed to dry out. After incubation at room temperature, the slides are rinsed free of excess labelled antiserum by immersion in PBS contained in a beaker. Further washing in PBS is carried out for one hour and with several changes of solution, by placing the slides in staining troughs suported on a horizontal shaker (Gallenkamp). Alternatively the slides can be placed in glass beakers containing stirring rods and supported on magnetic stirrers. Sections treated with blocking serum are further treated with FITC labelled sheep anti-rabbit immunoglobulins and the incubation and washing procedure repeated.

In the second approach, referred to as the 'sandwich' technique, the sections are treated with unlabelled sheep anti-rabbit immunoglobulin, incubated for half an hour and washed in PBS as previously described. At the same time controls are set up using unlabelled normal sheep serum. All sections are then further treated with rabbit FITC labelled anti-sheep immunoglobulin (Wellcome Research Laboratories), incubated for half an hour and washed in PBS. This method is more sensitive than the previous one in localizing the antigen (rabbit immunoglobulin) in the tissue, but may result in more 'non-specific fluorescence' than in the direct approach.

For viewing under the fluorescence microscope, the slides are wiped free of PBS save for a small area around the sections, and a drop of mounting medium (a mixture containing one part PBS and nine parts glycerine) is applied to the section, followed by a cover slip. We use a Wild M20 fluorescence microscope, which is provided with a lamp housing to accommodate Osram HBO 200 high pressure mercury burners as the source of exciting u.v. light. Two UGI exciter filters, a BG38 red absorbing filter and a GG13 barrier filter are used to provide a high and even transmission of fluorescent light and a minimum of exciting light. Under these conditions FITC conjugated antibodies that have reacted with antigen in the tissue display a bright apple green specific fluorescence.

Different areas of tissue may display a somewhat variable localization of rabbit immunoglobulins, but usually the yolk sac splanchnopleur has a distribution of specific fluorescence as shown in Plate 8.3a, b. Within the yolk sac endoderm rabbit immunoglobulin is contained within vesicles. Specific staining will also be observed within the basement membrane of the yolk sac endoderm and below this in the vascular mesenchyme and in the lumen of the vitelline vessels, indicating protein that has been transported from endodermal cells. The transport mechanism is an intracellular one and there is usually no evidence of any intercellular transport of proteins. Control sections should reveal only a dull green background auto-fluorescence, but erythrocytes within the vitelline vessels often show a much brighter non-specific greenish-yellow fluorescence. In the blocking test, there should be considerable reduction of specific fluorescence since unlabelled specific antibodies occupy the antigenic sites, thus preventing access of FITC labelled specific antibodies to the antigen.

Antibodies (γ G-globulin) and several other serum proteins (albumin and low molecular weight α – and β – globulins) are also transported to the exocoelomic and amniotic fluids. The major site of this transmission is the paraplacental chorion (Wild, 1970) and γ – globulin can readily be demonstrated in this membrane by using exactly the same procedures described for the yolk sac splanchnopleur. However, γ G – globulin has a different localization, as revealed by specific fluorescence, to that in the yolk sac endoderm. The paraplacental chorion is composed of bi- and multinucleate cytotrophoblast cells, which rest on an ill-defined basement membrane, underneath which is a loose network of collagen fibrils, fibroblasts and finally a mesothelium bordering the exocoel. Specific fluorescence is rarely found within the cytotrophoblast, but intercellularly, around the cells, and diffusely distributed throughout the loose mesenchymal tissue. The function of the proteins within the foetal fluids is not exactly known. They may be important as a source of smaller molecules and ions since many of these are know to bind to proteins. The amniotic fluid is swallowed by the rabbit foetus and the proteins it contains become highly concentrated within the gut (Wild, 1965). It is possible that such protein provides a rich source of amino acids for the foetus when digestion takes place.

Much interest has centred on the yolk sac splanchnopleur with respect to its ability to select different proteins during their transmission across it to the foeal blood. Such selection is a property of the endodermal cells, but the exact mechanism whereby this is brought about is still not known.

REFERENCES

Allen, E. (1922), The oestrous cycle in the mouse. *Am. J. Anat.*, **30** 297–371.

Amoroso, E. C. (1952), 'Placentation'. In: A. S. Parkes (Ed) *Marshall's Physiology of Reproduction*. Vol. 2 (3rd Ed) Longmans, Green & Co., London.

Austin, C. R. (1961), *The Mammalian Egg*. Blackwell Scientific Publications, Oxford.

Balinsky, B. I. (1965), *An Introduction to Embryology*. pp. 48–55, W. B. Saunders Company, London.

Brambell, F. W. R. (1958), 'The passive immunity of the young mammal', *Biol. Rev.*, **33**, 488–531.

Brambell, F. W. R. (1970), *The transmission of passive immunity from mother to young*. Frontiers of biology, Vol. 18, North-Holland, Amsterdam.

Bronson, F. H., Dagg, C. P. and Snell, G. D. (1966), 'Reproduction'. In: E. L. Green (Ed) *Biology of the Laboratory Mouse*. (2nd Ed) McGraw-Hill, New York.

Dalcq, A. M. (1957), *Introduction to General Embryology*, pp 103–28. Oxford University Press, London.

Daniel, J. C. (Ed) (1971), *Methods in Mammalian Embryology*, W. H. Freeman and Company, San Francisco.

Davies, J. (1958), 'The Physiology of Fetal Fluids', In: C. A. Villee (Ed) *Gestation, Transactions of the Fourth Conference*, 1957, Josiah Macy Jr. Foundation, New York.

Fowler, R. E. and Edwards, R. G. (1957), 'Induction of superovulation and pregnancy in mature mice by gonadotrophins', *J. Endocrinol.* **15**, 374–384.

Gardner, R. L. (1968), 'Mouse chimaeras obtained by the injection of cells into the blastocyst', *Nature, Lond.*, **220**, 596–597.

Graham, C. F. (1970), 'Parthenogenetic mouse blastocysts', *Nature, Lond.*, **226**, 165–167.

Hillman, N., Sherman, M. I. and Graham, C. F. (1972), 'The effect of spatial arrangement on cell determination during mouse development', *J. Embryol. exp. Morph.*, **28**, 263–278.

Marsden, H. M. and Bronson, F. H. (1964), 'Estrous synchrony in mice: alteration by exposure to male urine', *Science, N.Y.* **144**, 1469.

Mintz, B. (1962), 'Experimental study of the developing mammalian egg: Removal of the zona pellucida', *Science, N.Y.* **138**, 594–595.

Mintz, B. (1965), 'Experimental genetic mosaicism in the mouse'. In: G. E. W. Wolstenholme and M. O'Connor (Eds.) *Preimplantation Stages of Pregnancy.* Ciba Foundation Symposium. Churchill. London.

Mintz, B. (1971a), 'Clonal basis of mammalian differentiation'. In D. D. Davies and M. Balls (Eds) *Control Mechanisms of Growth and Differentiation.*, S. E. B. Symposia XXV. Cambridge University Press.

Mintz, B. (1971b), 'Allophenic mice of multi-embryo origin'. In: J. C. Daniel (Ed) *Methods in Mammalian Embryology.* W. H. Freeman and Company, San Francisco.

Mossman, H. W. (1937), 'Comparative morphogenesis of the foetal membranes and accessory uterine structures', *Contr. Embryol. Carneg. Instn.*, **26**, 129–246.

Mulnard, J. (1961), 'Problemes de structure et d'organisation morphogenetique de l'oeuf des Mamiferes', In: *Symposium on the Germ Cells and Earliest Stages of Development*, pp. 639–688. Fondazione A. Baselli, Instituto Lombardo, Milan.

Mystakowska, E. T. and Tarkowski, A. K. (1968), 'Observations on CBA – p/CBA – T6T6 mouse chimaeras', *J. Embryol. exp. Morph.*, **20**, 33–52.

Nairn, R. C. (1969), *Fluorescent Protein Tracing* (3rd Ed), E. & S. Livingstone Ltd., London.

Runner, M. N. and Palm, J. (1953), 'Transplantation and survival of unfertilised ova of the mouse in relation to postovulatory age', *J. exp. Zool.*, **124**, 303–316.

Scott, W. and Ray, P. M. (1972), 'Euthanasia'. In: 'The UFAW Handbook on the Care and Management of Laboratory Animals', Churchill Livingstone, Edinburgh.

Snell, G. D. and Stevens, L. C. (1966), 'Early embryology'. In: E. L. Green (Ed) *Biology of the Laboratory Mouse*. (2nd Ed) McGraw-Hill, New York.

Steptoe, P. C., Edwards, R. G. and Purdy, J. M. (1971), 'Human blastocysts grown in culture', *Nature, Lond.*, **229**, 132–133.

Stern, M. S. and Wilson, I. B. (1972), 'Experimental studies on the organisation of the preimplantation mouse embryo. 1. Fusion of asynchronously cleaving eggs', *J. Embryol. exp. Morph.*, **28**, 247–254.

Stern, M. S., (1972), 'Experimental studies on the organisation of the preimplantation mouse embryo', 11. Reaggregation of disaggregated mouse embryos', *J. Embryol. exp. Morph.*, **28**, 255–261.

Tarkowski, A. K. (1961), 'Mouse chimaeras developed from fused eggs', *Nature, Lond.*, **190**, 857–860.

Tarkowski, A. K., Witkowska, A. and Nowicka, J. (1970), 'Experimental parthenogenesis in the mouse', *Nature, Lond.*, **226**, 162–167.

Tarkowski, A. K. and Wroblewska, J. (1967), 'Development of blastomeres in mouse eggs isolated at the 4- and 8-cell stage', *J. Embryol. exp. Morph.*, **18**, 155–180.

Van der Lee, S. and Boot, L. M. (1956) 'Spontaneous pseudopregnancy in mice', *Acta. Physiol. Pharmacol. Neer.*, **5**, 312–315.

Whitten, W. K. (1957), 'Effect of exteroceptive factors on the oestrous cycle of mice', *Nature, Lond.*, **180**, 1436.

Whitten, W. K. (1971), 'Nutrient requirements for the culture of preimplantation embryos in vitro'. In: G. Raspe (Ed) *Advances in the Biosciences. No. 6. Schering Symposium on Intrinsic and Extrinsic Factors in Early Mammalian Development*, Pergamon, Oxford.

Whitten, W. K. and Biggers, J. D. (1968), 'Complete development *in vitro* of the preimplantation stages of the mouse in a simple chemical defined medium', *J. Reprod. Fert.*, **17**, 399–400.

Wild, A. E. (1965), 'Protein composition of the rabbit foetal fluids'. *Proc. roy. Soc. B.* **163**, 90–115.

Wild, A. E. (1970), 'Protein transmission across the rabbit foetal membranes', *J. Embryol. exp. Morph.* **24**, 313–330.

Wild, A. E. (1973a), 'Fluorescent protein tracing in the study of endocytosis'. In: J. T. Dingle (Ed) *Lysosomes in biology and pathology*, Vol. 3, North-Holland, Amsterdam.

Wild, A. E. (1973b), 'Transmission of immunoglobulins and other proteins from mother to young'. In: J. T. Dingle (Ed), *Lysosomes in biology and pathology*. Vol. 3, North-Holland, Amsterdam.

Wilson, I. B., Bolton, E. and Cuttler, R. H. (1972), 'Preimplantation differentiation in the mouse egg as revealed by microinjection of vital markers', *J. Embryol. exp. Morph.*, **27**, 467–479.

Subject Index

Numbers in italic refer to pages on which illustrations occur.

Aceto-carmine, for staining mouse eggs, 212
Acridine orange, 205–206, 212
Acrosome, 204, *205*
Adhesive gland, 103
  in ectodermal explants, 131, *Pl 6.2*
Albumen, effect on chick blastoderm, 175
  as bacteriostatic agent, 178
Albumin, bovine serum, 215
  in culture fluid, 215, 218
  of rabbit serum, 229
  as standard, 151
Allophenic mice, 214
Ambystoma mexicanum, see Axolotl
Ambystoma punctatum, 112
Agarose, 150
Allantois, of rabbit, 222, 226, *222, 223, Pl. 8.1*
  fluid of, 226
Amnion, of chick, 178
  of rabbit, *222, 223,* 226, *Pl. 8.1*
    fluid of, 222, 226, 229, 230
    folds of, 221, *221*
Amphibia, embryos, 96–97
  decapsulation of, 124–127
  experiments on, 119–122
Amplexus, 100
Ampulla, of mouse oviduct, 194, *195, 196, 198,* 209, *Pl. 8.2*
  of vas deferens, 203, 204
Anamorphic change, in crustacea, 75
Anatrapetic movement, in locust embryo, 59
Annelids, cleavage, 14
  development of, 36–41
  and molluscs, 29
Annual fish, 83–84
Anoestrus, 187
Anthracene blue, 171

Antibodies, detection of, 152, *153*
  against crystallins, 152
  in foetus and newly born, 223
  fluorescent labelled, 224, 226–229
  to protein antigens, 150
  from rabbits, 228
  to rabbit immunoglobulins, 227
  from sheep, 227, 228
  transmission of, 223–224
Antimetabolites, 178
Antiserum, fluorescent labelled, 158
  against lens crystallins, 152
  preparation of, 149–150
Anura, normal tables of, 97
Anus, in deuterostomia, 14
Aphyosemion scheeli, 85, 86–87, *88,* 89
Aphyosemion australe, 85
Archenteron, 14
  of *Xenopus*, 103, 130, *135*
  of *Echinus*, *21*
Area opaca, 169, 174, *177*
Area pellucida, 169, *174, 177*
Area vasculosa, *177*
Aristotle's lantern, 17
Artemia salina, development of, 76–77, *77*
  as food, 119
Artificial fertilization, of, sea urchins, 15
  Echinus, 17–20
  Ciona, 25
  Goldfish, 91–92
  Trout, 90–91
  Xenopus, 122–123
Ascaris lumbricoides, dissection, 43–44, *45*
  female reproductive system, 42–44
  fertilization, 44
  larvae, 44
  sources, 43

Ascaris megalocephala (*Parascaris equoruum*), 42
Ascidians, development of, 14–15, 20
 location, 22
 tadpole, 22, *26*
Ascidiella aspera, 22
Ascidiella scabra, 22
Asterias rubens, 15
Axolotl, breeding season, 97, 110, 111
 cleavage, 113, *115*
 development of, 112–118, *115*, *116*
  effect of temperature, *118*
 egg number, 110
 gastrulation, 114, *115*
 maintenance, 109–110
 neoteny, 142
 rearing of, 119
 removal of egg jelly coat, 125
 spawning, 110
 spermatophores, 110
 effect of thyroxine, 142
Azocarmine, 157

Baculum (Os penis), of mouse, 201, 202
Balbiani rings, see chromosome puffs, 72
Banding patterns, in polytene chromosomes, 68
Barnacles, 1, 76
Birds, see chick,
Bilaminar omphalopleur, of rabbit, 220, 222, *220–233*, 225
Biliverdin, in amniotic fluid, 226
Blastema, 149
Blastocoele, of *Echinus*, 21
 of mouse, 207, 214, Pl. 8.2
 of rabbit, 220
 of *Xenopus*, 102, 129, 130
  collapse of, 127, 133
  dorsal lip implants in, 134, *135*
Blastocyst, of mouse, 207, Pl. 8.2
 disaggregation of, 214
 in uterine lumen, 208, 209
 parthenogenetic development, 199
 removal from uterus, 210–211, *211*
 transfer to uterus, 213
 of rabbit, 220
Blastoderm, of annual fish, 84
 of *Aphyosemion*, 86, 87, *88*, 89
 of chick, 167–179, 221

explants of fish, 93
extra-embryonic, 87
formation in insects, 52
 of locust, 55, 59
 of quail, 179
 of teleost, 94
Blastokinetic movement, of insect embryos, 59
Blastomeres, of ascidian, 14
 destruction of, 30, 94
 loss of, 29
 of mouse, 207, 208
  culture of, 213, 214
 of sea urchin, 14
 separation of, 15
Blastopore, of amphibia, 133
 of Axolotl, 114
 in Deuterostomia, 14
Blastula, of, *Aphyosemion scheeli*, 86
 Ascidia, 14
 Axolotl, 113
 Echinoderm, 14
 Xenopus, 102
Blood islands, of chick, *177*, *178*, 179
Blowfly, see *Calliphora*
Brain, in *Aphyosemion scheeli*, 87
 vesicles of chick embryo, 169
Breeding, see also culture of animals,
 of, *Calliphora*, 61–62
  goldfish, 82
  killifish, 85
  Locusta, 54
  tropical fish, 81
Breeding season (cycle), Amphibia, 97
 Axolotl, 110
 Crepidula fornicata, 34
 goldfish, 90
 Littorina saxatilis, 34
 mammals, 187
 Patella species, 31
 Pomatoceros triqueter, 39
 trout, 90
 Tubifex, 36
Brine shrimps, see *Artemia*
Brood pouch, of *Daphnia*, 75
 of *Littorina saxatilis*, 35
Brood size, in *Lebistes*, 89
Bulbo-urethral (Cowper's) glands, of mouse, 203

Calliphora erythrocephala, 50, 53, 59, 61–67
 embryology of, 63

SUBJECT INDEX

larva of, 61–67, *62*, *66*
life cycle of, *62*
post-embryonic development of, 54
Cell lineage, in Annelids, 29, 36
   in Molluscs, 29, 30, 36
   in Nematodes, 29, 42
Cell movement, in chick blastoderm, 179
   in wound healing, 138
Cellular metaplasia, in Wolffian regeneration, 143
Centrifugation, of eggs, 30, 52
   of embryos, 52, 93
Centrioles, of mouse sperm, 206
Centromere, of salivary gland chromosome, 71
Cervix, of mouse, 196
Chick, blastoderm, 5, 167, 170
   culture of, 172–179, *176*
   infection of, 178
   outgrowth of, 177, 178, 179
   removal of, 168, *170*
   stained preparations, 171–172
   chorio-allantoic grafting in, 166, 180–183, *183*
   development of, 167–169
   eggs, 166
      incubation of, 167
      storage of, 167
      unfertilized, 171
   embryonic membranes, 166
Chimaeras, mouse, 214, 219
Chironomus, 68
   larvae, 70
   salivary gland chromosomes, 69
   puffing pattern in, 69
Chloroform, for killing mice, 196
   for killing insects, 55
Chordamesoderm, in amphibian exogastrulae, 127, 128
   in formation of neural plate, 127, 128
Chordates, 14
Chorio-allantoic grafting, 166, 180–183, *183*
Chorio-allantoic placenta, of mammals, 186
   of rabbit, 221, *222*, *223*, 225, *Pl. 8.1*
Chorio-allantois, of chick, 180, 181, 182, *183*
Chorion, of egg, *Aphyosemion scheeli*, *88*, 89
   Calliphora, 63

fish, 93, 94
insect, 51
locust, 56, *57*
killifish, 83, 85, 93
removal of, 57, *58*, 63
Chorion, paraplacental, of rabbit, 219, 221, *221*, *222*, 225, 226, 229, *Pl. 8.1*
Chorionic gonadotrophin, for induction of ovulation, 122
   for injection into *Xenopus*, 98, 99, 123
   for superovulation, 194–195
Chromocentre, of salivary gland chromosomes, *Pl. 4.1*
Chromosomes, and development, 54
   elimination of, 42
   lampbrush, 55
   maps of, 72
   in mouse,
      formation of pronucleus, 207
      meiosis, 206
      number of, 200
   polytene, 67, 68, 69, 71–74
   puffs, 69, 72–74
Ciliation, of ectodermal explants, 131, 132
Ciona intestinalis, breeding season, 23
   cleavage, 26
   description, 22
   development of, *26*
   dissection, 23, *24*
   fertilization, 25
   location, 22, 23
   metamorphosis, 26
   ovary, 23, *24*
   oviduct, *24*, 25
   tadpole, 25, *26*
   testes, *24*, 25
   vas deferens, *24*, 25
Cleavage, of eggs, annelids, 14
   axolotl, 113, *115*
   chicks, 167
   Ciona, 26
   insects, 52
   killifish, 86, *88*
   mammals, 213
   molluscs, 14
   radial, 14
   spiral, 14
   synchronous, 86
   Tubifex, 37, *38*
   Xenopus, 102

Clitorial glands, of mouse, 196, *197*
Clitorium, of mouse, 196
Coagulating glands, of mouse, *202*, 203, 209
Coleoptera, 63
Collagen, in yolk sac endoderm, 229
Collembolids, 52
Complan, as food, for *Artemia*, 76
  for *Xenopus*, 107, 109
Conceptus, 2
  of mouse, 186
  of rabbit, 224, 225, 226, Pl. 8.1
Copulation, in *Lebistes*, 90
Cornea, of amphibia, *136*
  culture *in vitro*, 148
  differentiation at metamorphosis, 146
  effect of thyroxine, 139
  regeneration of lens from, 142–149, Pl. 6.4
  structure of, 146, Pl. 6.4
  transplants of, 149
  Xenopus tadpole, *145*
Corpora allata, of diptera, 63, 64
Corpora cardiaca, of diptera, 64
Corpora lutea, of mouse, 189
Corpus uteri, of mouse, 196, *197*, 210
Cortex, egg, 30
  of insects, 51, 52
  of *Tubifex*, 37
Crab, larva of, 75
Crayfish, 75
Crepidula fornicata, 29, 30, 32
  breeding in, 33
  breeding season of, 34
  location of, 32
  oocyte of, 33
  oogenesis in, 33
  oogonia, of, 33
  seminal vesicle of, 34
  spiral cleavage in, 34
  veliger larva of, 34
Cruelty to Animals Act, 2, 3
  and mammals, 185
  and tadpoles, 3, 143
Crustacea, 50, 74–77
  anamorphic change in, 75
  development of, 75
  larva of, 75–76
  metamorphic change in, 75
  moulting in, 75
Crystallins, *see also* lens proteins,
  of chick lens, 159

classes of, 149, *158*, 157–159
determination of concentration, 151, 153
and lens differentiation, 159
immunoelectrophoretic analysis of, 149
preparation of, 150, 153
in regenerating lens, 159
sub-units of, 157
Culture of animals, *Ascaris*, 44
  Calliphora, 61–62
  Drosophila, 67–68
  insects, 54
  Locusta, 54
  Rhabditis, 46
Culture of embryos *in vitro*, chick, 172–179
  mouse, 215–219
Culture of tissues, cornea, 148
  ectodermal explants, 128–132
  isolated tadpole tails, 160–161
Cumulus oophorus, of mouse, 198, 199, 204, 209
Cuticle of insects, influence of type by hormones, 63, 72
  secretion by epidermis, 63, 72
  tanning of, 67
Cyprinodontidae, 83
Cypris, 76
Cysteine hydrochloride, activation of papain, 125–126
Cystoblast, in insect ovary, 51
Cytochrome oxidase, in *Tubifex*, 38
Cytoplasmic droplet, of spermatozoa of mouse, 204, *205*
Cytotrophoblast, of rabbit, 229

Daphnia, 75, 76
Decapsulation, of fish eggs, 93
  of *Xenopus* eggs,
    chemical, 125–127
    mechanical, 124–125
Decidua basalis, of rabbit, 225, Pl. 8.1
Dendrodoa grossularia, 22, 27
Dentalium, 30
Desoxyribonucleic acid (DNA), in insect epidermis, 67
  in polytene chromosomes, 69
  reaction with acridine orange, 206, 212
Deuterostomia, 14
Diapause, in annual fish, 84
Dictyate stage, of mouse egg, 193

SUBJECT INDEX

Dioestrus, in mice, 187, 189–190, 192
Diplotene, of mouse egg, 193
Diptera, 52
    pupation in, 61, 63
    salivary glands of, 68
Dopa-decarboxylase, in *Calliphora* larva, 67
Dorsal lip, of amphibian gastrula, of axolotl, 114
    grafts of, 133, 140
    implantation of, 133–134, *135*
    of *Xenopus*, 102, 130
Double embryos, of *Tubifex*, 39
Drosophila, 50, 52, 53, 67–74
    culture of, 67–68
    larvae, 54, 67–74
    life cycle of, 67
    salivary glands, 68, *Pl. 4.2*
        chromosomes of, 69, *Pl. 4.4, Pl. 4.3*

Ecdysial glands, *see* prothoracic glands
Ecdysone, 63, 65, 66, 67, 69, 70, 72, 73
Echinoderms, 14, 15
    cleavage of, 14
    development of, 14–15
    hybridization of, 19
Echinus acuteus, 16
Echinus esculentus, 15
    artificial fertilization in, 15, 19–20
    breeding season in, 16
    description of, 16, 17, *18*
    development of, 20, *21*
    gastrulation in, 20
    gonads of, 18
    induced spawning in, 17–20, 19
    location of, 15, 16
    ovaries of, 19
    storage of, 17
    testes of, 19
Ectoderm, amphibian, differentiation in explants, 129
    in exogastrulae, 127
    mouse, 208
    rabbit, 221
Ectodermal explants, of *Xenopus* gastrulae, 128–132
    differentiation of, 131–132, *Pl. 6.2*
    method of preparation, 129–130, *130*
    survival of, 131, *132*
Egg, cortex, 37, 51, 52

membranes, 2
masses, 75
mosaic, 52
parthenogenetic activation of, 15
pigment, 40, 101
pole plasms, 30, 37, 39, 52
regulative, 52
sacs, 77
shell, crustacean, 75
Egg(s) of, *Artemia*, 76
    ascidians, 14
    axolotl, 112
    Calliphora, 61, 62, 63
    chick, 166
    Drosophila, 68
    echinoderms, 14
    fish, 92
    human, 213
    insects, 51–52
    killifish, 83
    mammals, 12, 193
    mouse, activation of, 206
        culture of, 212–219
        fusion of, 214, 218–219
        isolation of, 197, 209–210
        staining of, 212
    trout, 90
    Xenopus, 5, 101
Egg, number, in, axolotls, 111
    Calliphora, 62
    goldfish, 91
    Locusta, 54, 56
    Xenopus, 100
Ejaculation, in mice, 203
Electrolytic method of sharpening needles, 8–9
Electrophoresis, of lens protein, 155, *156*
Embryonic axis, of *Aphyosemion scheeli*, 87
    induction of, 140
Embryonic membranes, 6
    of chick, 166
Embryonic plate, of mouse, 220, 221
Endocrine glands, of insects, corpora allata, 63
    corpora cardiaca, 64
    prothoracic glands, 64
    Weismann's ring, *64*
Endocytosis, of proteins, 222, 224
Endoderm, amphibian, in exogastrulae, 127
    mouse, 208

rabbit, 220, 221
  of yolk sac splanchnopleure, 222, 223, 225
Endometrial glands, of mouse, during oestrous cycle, 188, 189
Epidermis cells, in *Calliphora*, 65, 66, 67
  in *Drosophila* larvae, 72
  of insects, 63, 65
  in lens formation, 135, *136*, 137
Epididymis, of mouse,
  caput, 198, *202*, 203
  cauda, 198, *202*, 203
  corpus, *202*, 203
  ductus, 203
  location of sperm in, 204
Epigenetic factors, in mouse development, 213
Exocoel, of rabbit, 221, 222, *220–223*, Pl. 8.1
  fluid of, 222, 229
  mesothelium of, 229
Exogastrulae, of *Xenopus*, 127, 128
Exogastrulation, in amphibian embryos, 127, 129
  in *Xenopus* embryos, 127
Explants, of chick blastoderm, 172–178, *177*
  of fish blastoderm, 93
  ectodermal, from ampihbian gastrulae, 128–132
Eye, vertebrate (amphibian), development of, 134–135, *136*
  lens of, 135, 136
  optic nerve in, 134
  pigment coat of, 134, 135
  sensory retina of, 134, *136*

Feeding, of *Xenopus* tadpoles, 107, 108
  of axolotl larvae, 119
Fertilization, *Ascaris*, 44
  Ciona, 25
  Echinus, 19–20
  insects, 52
  killifish, 86
  Lebistes, 89
  mouse, 194, 206–207
  Patella, 31, 32
  Pomatoceros, 39–40
  trout, 82
Fertility, of chick eggs, 166
Fibroblasts, in rabbit chorion, 229

Fish, 76, 81–94
  larvae of, 76
Fluorescein isothiocynate, 227
Foetal membranes, of mammals, 186, 219
  chorion, 219
  yolk sac splanchnopleur, 219
  of rabbit, arrangement and development of, 220–223, Pl. 8.1
Follicle, of mouse ovary, atretic, 193
  primary, 193
    control of growth, 193
    secretion of oestrogen by, 193
Follicle cells, ovarian, of insects, 51
  of goldfish, *84*
  mammalian, 193
  of mouse, 198, Pl. 8.2
Follicle stimulating hormone (FSH), 193, 194
Folin and Ciocalteu's reagent, 151
Foot pad cells, of *Sarcophaga*, 74
Forebrain, of chick, *177*
Freund's adjuvant, 152
Frogs, 1
Fundulus heteroclitus, 86, 93, 94
Fungal infection, of killifish, 85
  of axolotls, 110, 119

Gametogenesis, in teleosts, 82
Gastrula, of *Aphyosemion scheeli*, 86
  of ascidians, 14
  of echinoderms, 14
  of *Xenopus*, *104*, 102–103
Gastrulation, in ascidians, 22
  in Axolotls, 114, *115*, 127
  of *Echinus*, 20
  in rabbit, 220
  significance in vertebrate embryos, 133
  in *Xenopus*, 101, 102–103, *104*, 105
Genes, banding patterns and, 69
  derepression of, 67, 70
  ecdysone and, 70
  effect on development, 50
  and lens differentiation, 149
  transmission by sperm, 200
Genetic engineering, 214
Genital pores, of *Echinus*, 19
Germ bands, of *Tubifex*, 37, *38*, 39
Germ cells, of insects, 52
Germinal vesicle, of goldfish, *84*
  mammalian, 193

SUBJECT INDEX

of *Patella*, 32, *33*
of *Pomatoceros*, 40, *41*
Gestation, length in rabbit, 224
Glans penis, of mouse, 201
Glassware, 5
Globulins, see also immunoglobulins,
    α–, 229
    β–, 229
    γ–(G), 223, 224, 229
Goldfish, 81, 82–83
    breeding of, 82
    dissection, 82
    ovary, *83*
Golgi apparatus, in mouse spermatids, 204, 205
Gonadotrophin, 187, 193
Gonads, *Ciona*, 24
    Dendrodoa, 27
    Echinus, 18, *19*
    goldfish, 81, 82
    Patella, 31
    Pomatoceros, 40
Gonopodium, of *Lebistes*, 90
Growth, of *Artemia*, 76
    of insects, 53
    of *Xenopus* tadpoles, 109
Guppy, see *Lebistes*
Gynogenetic haploids, of amphibia, 124

Hair loops, preparation of, 9, *10*
    use in removing optic vesicles, 137–138, *138*
Hatching, of *Artemia*, 76
    of axolotls, 117
    of killifish, 86
    of *Xenopus*, 106
Head, fold, 167, 173, *174*, 177
    organizer, 134
    process, 173, *174*
Heart, embryonic, of *Aphyosemion scheeli*, 87
    of chick, 169, *177*, 178, 180
Hermaphrodites, in mice, 214
Hemimetabolous insects, 53
Hindbrain, of chick, *177*
Holometabolous insects, 53, 63
Holtfreter's solution, 120–121
Home Office, and Cruelty to Animals Act, 2
    Inspector, 2
    licence, 2
    and amphibian larvae, 142, 143
    and vertebrate larvae, 2
Honey-bee, 52
Hoppers, of *Locusta*, 54, 59, *53*
Hormones, in amphibian metamorphosis, 159–160
    in *Calliphora*, 54
    control of pupation, 61
    and reproductive cycle of *Lebistes*, 90
    in mammalian reproduction, 185, 187
    effect on mammalian testes, 200
Human development, on film loops, 186–187
Hyaluronidase, for removal of cumulus oophorus, 198, 199
    and sperm penetration, 204
Hybrid vigour, in mice, 215
Hybridization, in sea urchins, 16
Hydropyle, of locust egg, *57*
Hymenoptera, 52
Hypothalamus, in mammalian reproduction, 187
Hydroides, 39

Ilyanassa, 29, 30
Imago, of *Drosophila*, 67
    of holometabolous insects, 63
    of locust, 59
Immunity, passive, 223
    transmission of, 223
Immunoelectrophoresis, analysis of lens proteins, 149–159
    slides, preparation of, 154–155
    staining of, 156–157
    template for, *154*
Immunofluorescence, for detection of crystallins, 158–159
    for detection of γ-globulins, 224, 227–229
Immunoglobulins, antibodies against, 227
    in rabbit foetal membranes, 185
Implantation, of blastocyst, in mouse, 186, 199, 208
    in rabbit, 220
Implantation, of dorsal lip, 133, *135*
Incubators, 4, 5
Induced spawning, in *Echinus*, 17–20, *19*
    in *Psammechinus*, 18, 19
Inductive interaction, and formation of nervous system, 129

in lens formation, 135
primary, analysis of, 133
in *Xenopus* gastrulae, 103
Inner cell mass, of mammalian
    embryo, 213
  mouse, 207, 208, 214, *Pl. 8.2*
  rabbit, 220
Insects, development of, 50–54, *60*
  hemimetabolous, 53
  holometabolous, 53
Instars, of *Artemia*, 77
  of *Calliphora*, 64–65, 66
  of *Drosophila*, 67, 68, *69*, 70
  of *Locusta*, 53, 54,
Instruments, 6–12
Ionagar, as agar base, 137
  for immunoelectrophoresis, 150, 154
Iris, and lens regeneration, 142, 159

Janus green, 206
Jelly coat, of eggs, of *Patella*, 32
  of *Xenopus*, 100, 123
  removal of,
    from axolotl, 112, 125
    from *Xenopus*, 124–127
Juvenile hormone, in insects, 63

Killifish, 81, 83–85, 93

Laboratory, 3
  temperature of, 4
Lacto-propionic acid orcein stain, 71
Lampbrush chromosomes, in the
  locust, 55
Larvae, amphibian, 96
  experiments on, 141–162
  and Home Office requirements, 142
  lens crystallins in, 153, 157, *158*
  metamorphosis of, 142
  regeneration in, 142
  of *Artemia*, 76
  of *Ascaris*, 44
  ascidian, 14, 27
  of *Calliphora*, 59, *62*, 61–67
  of *Chironomus*, 70
  crab, 75
  crustacean, 75–76
  cypris, 76
  of *Dendrodoa*, 28
  of *Drosophila*, 54, 67–74
  insect, 50, 53

metanauplius, 76, *77*
nauplius, 75
of *Rhabditis*, 40
sea urchin, 14
trochophore, 29, 32, *33*
urodele, 117, 141
veliger, 32, *33*, 34
vertebrate, 2
zooea, 75
Lebistes reticulatus, 87, 89–90, *91*
'Lee Boot' effect, 192
Legal considerations, 2
Lens, analysis of proteins in, 144, 149–159
  in *Aphyosemion scheeli*, 87
  crystallins of, 149
    during regeneration, 159
  determination of protein content, 151
  epithelium of, *136*, 147, *Pl. 6.4*
  fibres, *136*, 147
  formation of,
    from cornea, 148–149
    in vertebrates, 135, *136*
    in amphibians, *136*
  origin from ectoderm, 143
  placode, *136*
  regeneration from cornea, 142–149, *Pl.6.4*
  regeneration from iris, 142
  removal of,
    and Cruelty to Animals Act, 3
    in *Xenopus*, 144, *145*, 150, 153
  vesicle, *136*
Lensectomy, *see* lens, removal of
Lentoid, 147
Lepidoptera, 63
Leptotene, in mammalian oocyte, 193
Life cycle, life history, of *Artemia*, 76, 77
  of *Calliphora*, 61–62, *62*
  of Crustacea, 75–76
  of *Drosophila*, 67
  of *Locusta*, 53, 54
Ligaturing, of eggs and embryos, 52
  of larvae,
    Calliphora, 64–67, *66*
    Drosophila, 72–74, *74*
Limb, development in urodeles, 117
Limb bud, chick, isolation, 181
  grafting, 180–182, *183*
Limnaea stagnalis, 30
Limpet, *see Patella*

SUBJECT INDEX

Lithium, 30
Littorina, littoralis, 34
  littorea, 34
  neritoides, 34
  saxatilis (*rudis*), 30, 34–35
    occurrence, 34
    breeding season, 34
    brood pouch, 35
    removal from shell, 35, *35*
Locust (*Locusta*), African migratory, 54
  blastoderm, 59
  development of, 55–59, *60*
  egg pods, 55
  female reproductive system, *56*
  hopper, 54
  L. migratoria migratoiorodes, 54
  life cycle of, *53,* 54
  ovarioles, *51,* 55
  panoistic ovary, 54, 55
Luteinizing hormone (LH), and oestrous cycle, 193, 194
Lysosomal enzymes, in tail regression, 162
Lysozyme, in egg albumen, 178

Macromeres, in *Tubifex,* 37
Mammals, eggs of, see also ovum (ova), 193
  reproductive systems of, 185
Marine invertebrates, 5
Mating, of *Artemia,* 76, 77
  of mice, 208–209
  of rabbits, 224
Meiosis, in foetal ovary, 193
  in mammalian testes, 200
  in mouse egg, 199, 206
Melanocytes, of killifish, 87
Mesenchyme, of *Echinus* embryos, 21
  of allantois, 222
  of choirion, 230
Mesoblasts, of *Tubifex,* 39
Mesoderm, chick, *177*
  mouse, 208
  rabbit, 221
  Xenopus, 127, 135
Mesometrium, of mouse, 196, *197,* 210
Methylene blue, as vital dye, 140
Metamorphic climax, in *Xenopus, 108*
Metamorphosis, 26
  amphibian, 96, 142
    morphological changes, 159–160
    biochemical changes, 159–160

insect, hormonal control of, 53
  in *Drosophila,* 69, 72
  in *Calliphora,* 63–67, 90
  of *Pomatoceros,* 40
  of *Xenopus,* 107, *108,* 109, 139
    and differentiation of cornea, 146
    and differentiation of crystallins, 157
    premature, 142, 159
    retardation of, 142
    tail regression *in vitro,* 160–162
Metanauplius, of *Artemia,* 76, 77
Metoestrus, in mouse, 187, 189
  sub-division of, 192
Messenger RNA (mRNA), in insect epidermis, 67, 69
Micromeres, of *Tubifex,* 37
Micropylar zone, of locust egg, 56, 57, 59
Micropyle, of insect egg, 51
Microscope, 5
  fluorescence, 158, 185, 205, 228
Microworms, as source of food, 119
Midbrain, of chick, *177*
Millions fish, see also *Lebistes,* 87
Milt, trout, 90
Mitochondria,
  in molluscan egg, 38
  in spermatozoa, 206
Molluscs, cell lineage, 30
  development of, 29–30
Morphogenetic movements, in teleosts, 34
Morula, of *Pomatoceros,* 40
  of mouse, 207, 209, 211, Pl. 8.2
Mosaic, development, 30, 42
Mosaic eggs, 29, 36, 52
Mouse, blastocyst, 207, 209, 213, Pl. 8.2
  culture of, 215–219
  eggs, 206
    culture of, 215–219
  early development of, 186
  embryos of, 205–219, Pl. 8.2
  female reproductive system, 196–197, *197*
  male reproductive system, 201–204, *202*
  oestrous cycle of, 187–190, 192–193
  reproduction of, 186
  spermatozoa, 204–206, 208, 209
  timing of matings, 208–209
  vaginal smears from, 190–192

Moulting, of *Calliphora* larvae, 65
  of *Drosophila* larvae, 67, 68, 70
    prepupae, 68
  in crustacea, 75
  in insects, 53
    holometabolous, 63
  of *Locusta*, 68
Mouth pipette, 12, 198, *211*
MS 222, for anaesthetizing,
    amphibian embryos, 138
  amphibian larvae, 142, 144, 160
  for killing, adult *Xenopus*, 150
    chick embryos, 169
    fish, 82, 89
    Xenopus tadpoles, 153

N-acetyl dopamine quinone, in
    Calliphora larvae, 67
Nauplius, of crustacea, 75, 76
Needles, glass, preparation of, 7
  metal, preparation of, 7
    sharpening by sodium nitrate, 8
    sharpening electrolytically, 8, 9
Nematodes, 29, 41–42
  spiral cleavage in, 29
Nembutal, for killing rabbits, 152, 224
  for killing *Xenopus*, 123
Neoteny, in axolotls, 142
Neotenin, *see* juvenile hormone
Nettle powder, 107
Neural crest, of amphibia, 141
Neural folds, in *Xenopus*, 102, *104*, 114, *116*
Neural groove, of rabbit, *220*
Neural plate, formation of, in
    axolotls, 114
  in vertebrate embryos, 129
  in *Xenopus*, 103, *104*
  re-orientation of, 141
Neural tissue, relation to
    chordamesoderm, 128
  in extodermal explants, 131, *Pl. 6.2*
Neural tube, in axolotls, 114
  in chick, 169
  in *Xenopus*, 103, 105
Neurosecretory cells, in insect brain, 63
Neurulation, in amphibians, 101
  in axolotls, 114
  in ectodermal explants, 131
Neutral red, as vital dye, 140

New's culture method, for chick
    blastoderms, 172–180, *176*
  for quail blastoderms, *179*, 179–180
Newts, 1
Nile blue sulphate, as vital dye, 137
Nipagin, 68
Normal tables of development,
    amphibian, 97
  Xenopus, 105
  urodele, 117
Nuclear transplantation, in amphibia, 96, 141
Nuclei, of salivary gland cells, 70
Nucleoli, of goldfish oocytes, 83, *84*
  and lens differentiation, 147, *Pl. 6.4*
Nurse cells, in insect ovary, 51

Oestrous cycle, in mammals, 187
  in mouse, characteristics of stages, 188–189
    determination of length, 192
    effect of crowding, 192
    effect of male urine, 193
    length of stages, 188–189
    synchrony of, 193
  in rat, 187
Oestrus, in the mouse, 187, 188
Oestrogen, effect on pituitary, 193–194
  effect on vaginal epithelium, 188
  levels in *Lebistes*, 90
  secretion by follicle, 193
Oniscus, 75
Oocytes, of *Crepidula*, 33
  of goldfish, 83
  of insects, 51
  in mammals, 193
  pre-vitellogenic, *84*
  vitellogenic, *84*
Oogonia, in *Crepidula*, 33
  in mammals, 193
Oogenesis, in *Crepidula*, 34
  in goldfish, 83
  in *Locusta*, 55
  in mouse, 200
Optic cups, formation in vertebrates, 134, *136*
  of killifish, 87
Optic vesicle(s), of axolotl, 117
  of chick, 169, *177*, 180
  formation of, 134, *136*
  inductive influence of, 137, 143
  and metamorphosis, 160

SUBJECT INDEX

regulation of pieces of, 139
removal of, 137, *138*
transplantation of, in *Xenopus*, 134–140, *Pl. 6.3*
Otic pit, of axolotl, 117
  of chick, *177*
  of *Xenopus*, 106
Ovarian capsule, of mouse, 196
Ovarian cavity, of *Lebistes*, 89
Ovarioles, types of, in insects, 51
  of *Locusta*, 55
Ovary, ovaries, *Ascaris*, 44, *45*
  Ciona, 23, *24*
  Echinus, 19
  goldfish, 82, 83, *83*, 84
  insects, 51
  Lebistes, 89
  Locusta, 54, 55, *56*
  mammalian,
    during oestrous cycle, 188–189
    foetal, 193
    secretion of hormones by, 187
  mouse, 196, *197*
  Patella, 31
  Pomatoceros, 40
  Rhabditis, 47
  Tubifex, 36
Oviduct(s), of *Ascaris*, 44, *45*
  of *Ciona*, *24*, 25
  of goldfish, 82, *83*
  of *Locusta*, 55, *56*
  of mouse, 196, *197*, *198*, *Pl. 8.2*
    ampulla of, 194, 197, *198*, 209
    passage of eggs along, 207
    presence of eggs in, 209
    removal of eggs from, 209–210, 216
Oviposition, in *Drosophila*, 67
  in *Locusta*, 55
Ovo-viviparous forms, *Dendrodoa grossularia*, 27
  Littorina saxatilis, 34
  Rhabditis pellio, 44
  Tse-tse fly, 51
Ovulation, in mammals, 187
  induction in mouse, 194–195
  during oestrus, 188
  in rabbit, 224
  in *Xenopus*, 98
Ovum (ova), in mammals, liberation of, 193–194
  in mouse, 198, 205
    liberation of, 196

observation of, 199
Oysters, 1

Pachypanchax playfairii, 85
Pachytene, in mammalian oocyte, 193
Pannett-Compton solution, 173
Paracentrotus lividus, 16
Paraxial mesoderm, in chick, *177*
p-hydroxyphenyl pyruvic acid, 66
Parthenogenetic activation, 15
  of mouse eggs, 199
  of *Xenopus* eggs, 124
Papain, for decapsulating *Xenopus* eggs, 125
Patella, 30–32
  spiral cleavage in, 29
  aspera (*athletica*), 31–32
  intermedia (*depressa*), 31–32
  vulgata, breeding season in, 31
    development of, *33*
    diagnostic characters, 31
    fertilization of, 31, 32
    germinal vesicle of, 32, *33*
    location of, 30
    ovary of, 31
    polar bodies in, 32
    spiral cleavage in, 32
    testis of, 31
    trochophore of, 32, *33*
    veliger of, 32, *33*
Penis, of *Crepidula*, 34
  of mouse, 201, *202*, 203
Perivitelline space, of mouse egg, 206, 210, *Pl. 8.2*
Phaeodactylum, 40
Pinocytosis, 224
Pipettes, 9, 10, *11*
  mouth, 12, *211*, *217*, 218
Pituitary gland, and control of oestrous cycle, 187
  effect of oestrogen on, 193–194
  in metamorphosis, 159
  secretion of gonadotrophin by, 193
Placenta, of mammals, chorioallantoic, 186, 219
  definition of, 219
Plankton, 75, 76
Polar body (bodies), of mouse, 199, 206, 207, 210, *Pl. 8.2*
  of *Patella*, 32
  of *Pomatoceros*, 40, 41
  of *Tubifex*, 37, *38*
Polar lobes, of *Tubifex*, 37, *38*

Pole plasms, in insects, 52
  in molluscs, 52
  in *Tubifex*, 37, 39
Polyembryony, in parasitic Hymenoptera, 52
Polymorphonuclear leucocytes, during oestrous cycle, 188, 189, 190
Polyspermy, block to, 207
Polytene chromosomes, 67
  appearance of, *Pl. 4.1*
  in foot pad cells, 74
  puffing patterns in, 68–70, *Pl. 4.3*
Pomatoceros triqueter, 1, 39–41
  development of, *41*
  breeding season of, 39
  gonads of, 40
  metamorphosis in, 40
  spiral cleavage in, 29
  occurrence of, 39
Precipitin lines, 150, 154–157, *158*
Predetermination, 213
Pregnancy, 187
  and corpora lutea, 189
Prepupa, of *Drosophila*, 70, 72, 73
  moult of, 68, 74
Preputial glands, of mouse, 201, *202*, 203
Presumptive areas, of amphibian blastula, 140
Primitive streak, of chick, 169, 170, 173, *174*, 177
Pregnant mare's serum, 194
Prepuce, of mouse, 201, *202*
Proestrus, in mouse, 187, 188
Progesterone, 189
Pronase, 93, 218
Pronuclei, of mouse egg, 207, *Pl. 8.2*
Prometamorphic tadpoles, *108*, 160
Prophase, in mammalian oocyte, 193
Pronephros, of Axolotls, 117
  of *Xenopus*, 107
Propylthiouracil, effect on metamorphosis, 142
Prostaglandins, 203
Prostate gland, of mouse, *202*, 203
Protandrous hermaphroditism, 34
Protein(s), effect on synthesis by thyroxine, 162
  in foetal fluids, 230
  transport and selection by yolk sac splanchnopleur, 229, 230
Prothoracic glands, of insects, 63, 64
Protostomia, 14

Psammechinus miliaris, 15–16
  location of, 16
  induction of spawning in, 18
Pseudopregnancy, 187, 189, 192
Puffing patterns, in polytene chromosomes, 68–74
Pupa, of *Calliphora*, *62*, 64–67
  of insects, 53
  holometabolous, 63
Pupation, in *Calliphora*, 62, 65
  in *Chironomus*, 70
  in *Drosophila*, 67, 73
  of holometabolous insects, 63
  hormonal control of, 61
Puromycin, 162

Quail, culture of blastoderm, 179–180

Rabbit, antibodies from, 228
  blastocyst of, 220
  chorio-allantoic placenta, 219–224, *220–223*
  foetal membranes of, 186, 219–226, *220–223*, *Pl. 8.1*
  localization of immunoglobulins in yolk sac splanchnopleur, 229
  mating of, 224
  ovulation in, 224
  preparation of antisera from, 150, 152
  transmission of antibodies in, 224
Rana, 97
  sylvatica, 140
Rat, oestrous cycle of, 187
Regeneration, in amphibia, 96
  in urodele larvae, 142
Regulation, of eggs and embryos annelids, 14
  ascidians, 14
  echinoderms, 14
  insects, 52
  molluscs, 14
  teleosts, 94
Reproductive behaviour, in *Lebistes*, 90
Reproductive cycle(s), 1
  goldfish, 82
  Lebistes, 89, 90, *91*
Reproductive system, of female *Ascaris*, 42–44, *45*
  of female locust, 55, *56*
  of female mouse, 196–197, *197*

SUBJECT INDEX

of male mouse, 201–204, *202*
Reproductive tract, of mouse, during oestrous cycle, 188–190
Rete testis, of mouse, 203
Rhabditis, 42, 44, 46–48
   anomala, 46
   pellio, *47*, 48
      culture of, 46
      development of, 46, 47
      larvae of, 46
   terrestis, 46
Ring test, for detecting antibodies, 152, *153*
Ribonucleic acid (RNA), effect of throxine on synthesis of, 162
   staining with acridine orange, 212

Sacculina, 75
Saline, Pannett-Compton, 57, 173
   phosphate buffered, 227
Salivary glands, of Diptera, 68
   of *Drosophila*, *69*, Pl. 4.2
      preparation of, 70–71
      staining of, 71
Sarcophaga bullata, 74
Schistocerca gregaria, development of, 57, *60*, 61
Sea squirts, *see* ascidians
Sea urchins, *see Echinus* and *Psammechinus*
Seminal fluid, of mouse, 201
Seminal receptacle, of *Ascaris*, 44
Seminal vesicle, of *Crepidula*, 34
   of mouse, 201, *202*, 203, 209
Seminiferous tubules, of mouse, 200, 203
Serpula, 39
Sheep, antibodies from, 227, 228
   serum of, 228
Sinus terminalis, 225
Sodium hypochlorite, for dechorionating insect eggs, 57, 63
Sodium nitrate, for sharpening tungsten needles, 8, 9
Sodium pentobarbitone, see nembutal
Sodium thioglycollate, for decapsulating amphibian eggs, 126
Somatoblasts, 39
Somites, of *Aphyosemion scheeli*, 87, *88*, 89
   of axolotl, *116*, 117

of chick, 167–168, 169, *177*, 178
Sources of animals, *Artemia*, 76
   Ascaris, 43
   Calliphora larvae, 65
   Lebistes, 87
South African clawed toad, *see Xenopus*
Spawning, in axolotls, 110
   induction in *Xenopus*, 98–101
Sperm, amoeboid, 44
   motility in mice, 203
   preparation from mice, 198
   preparation from *Xenopus*, 123
   trout, 90
Spermatids, of mouse, 200, 204
Spermatophores, of axolotl, 110, 111
Spermatocytes, of mouse, 200
Spermatozoa, of mammals, 200
   of mouse,
      in ampulla, 209
      centrioles in, 206
      cytoplasmic droplet of, 204, *205*
      maturation of, 204
      mitochondria in, 206
      structure of, 204, *205*, 206
Spermateleosis, in mouse, 204, 205
Spermatogenesis, in *Crepidula*, 34
   in goldfish, 83
   in mammals, 200
Spermatogonia, of *Crepidula*, 33
   of mammals, 200
Spiralian egg, 36
Spiral cleavage, in, Annelids, 14
   Crepidula, 34
   molluscs, 14
   nematodes, 29
   Patella, 29, 32
   Pomatoceros, 29
Spratt's culture method, for chick embryos, 179
Stage numbering of embryos, amphibian, 101
   axolotl, 112
Starfish, *see Asterias rubens*
Steinberg's solution, 120–121
Stone loach, 93
Streptomyces gracilis, 218
Styela clava, 22
Sulphadiazine, used in culture media, 129, 139, 160
Superovulation, of mice, 194–195, 199

Tadpole, of ascidian, 22, *26*

and Cruelty to Animals Act, 143
isolated tails of, 159
of *Xenopus*, 106, 107, *108*
Tadpole tail, effect of thyroxine, 139
as graft site, 139
Tail bud, as graft site, 138
Tail organizer, 134
Teleosts, blastoderm of, 94
development of, 83, 94
gametogenesis in, 82
Temperature, effect on development, axolotls, 118
gastrulation, 105
Xenopus, 105
Teratogenic agents, 178
Testis, testes, of, *Ciona*, 24, 25
Echinus, 19
goldfish, 82
mammals,
effect of hormones on, 200
mouse, 201, *202*, 203
Patella, 31
Pomatoceros, 40
Rhabditis, 47
Testosterone propionate, 90
Tetramin, 85
Theca interna, production of oestrogen by, 193
Thyroidectomy, 159
Thyroxine, and metamorphosis, 139, 142, 159–160
and protein synthesis, 162
and RNA synthesis, 162
effect on tadpole tail, 159–162, *162*
Toluidine blue, for staining explants, 122
for staining mouse eggs, 212
Tooth carps, see Cyprinodontidae
Tri-iodothyronine, and metamorphosis, 142, 159
Triturus, 97
lens regeneration in, 142–143
Trophoblast, of mouse, 208, 213, *Pl. 8.2*
of rabbit, 220, *220*, 221, 222
Tropical fish, 1, 81
Trophectoderm, of mouse blastocyst, 208
Thymidine H^3, labelling of mouse eggs, 214
Trout, 1, 81, 82
artificial fertilization, 90
breeding season, 90

Tse-tse fly, 51
Trypsin, for decapsulating fish embryos, 93
Tubifex, cleavage in, 37, *38*
cocoons of, 36, 37
development of, 37, *38*
egg,
cortex of, 37
polar bodies of, 37, *38*
polar lobes of, 37, *38*
pole plasms of, 37
as food, 85, 86, 98, 109, 119
Tungsten needles, 8, 9, 11
Tunica albuginea, of mouse testes, 203
Tyrosine, and epidermal cells of *Calliphora*, 66

UFAW handbook, 3, 12, 13, 231
Umbilical cord, *223*, 225, *Pl. 8.1*
Urodeles, embryos, dorsal lip transplants in, 141
normal tables for, 97, 117
larvae,
regeneration of appendages, 142
regeneration of lens, 142–143
Uterus, uteri, of *Ascaris*, 43, 44, *45*
of mouse,
blastocysts from, 210–211, *211*
blastocysts in, 209
contractions of, 203
epithelium of, 208, *Pl. 8.2*
horns of, 196, *197*, 210, *211*
implantation in, 213
lumen of, 208
during oestrous cycle, 188–190
preparation of sections, 186
of rabbit, epithelium of, 220, 223, *223*
fluid of, 223
horns of, 225
lumen of, 220, *220*, 223, *223*, 224, 225
removal of conceptus from, 225

Vaginal plug, of mouse, 208–209
Vagina, during oestrous cycle, 188, 195
Vaginal smears, from mice, 188–192
changes during oestrous cycle, 188–190
Vas deferens, of *Ciona*, 24, 25
of goldfish, 82

SUBJECT INDEX

of mouse, *202*, 203, 204
Vasa efferentia, of mouse, 203
Veliger, of *Patella*, 32, *33*
  of *Crepidula*, *34*
Vesiculase, from mouse seminal vesicle, 202
Viruses, culture on chorio-allantois, 180
Vital staining, 94, 140
Vitelline artery, of rabbit, 225
Vitelline circulation, in *Aphyosemion scheeli*, 87
  in rabbit, 21, 224, 225
Vitelline membrane, of chick, 168, *170*, 173–175
  penetration of, in mouse, 205
  of quail, 179
  removal of, in *Xenopus*, 124
Vitelline vein, of chick, *177*
  of rabbit, 226
Vitelline vessels, of rabbit, 225, 229, Pl. 8.1, Pl. 8.3
Vitellus, of mouse, 199, 206, 207, Pl. 8.2
Viviparity, 81
  in *Lebistes*, 89

Watchmaker's forceps, 6, 7, *7*
Weismann's ring, *64*, 65, 66, 73, 74
White Leghorn, eggs of, 166
White worms, as food, 85, 86
'Whitten effect', 193
Whitten's medium, 215
Wolffian regeneration, 143
Woodlouse, *see Oniscus*
Wound healing, and cell movement, 138

Xenopus (*laevis*), cloacal papilla of, 10, 122
  development of, 101–109, *104*, *107*, *108*

ectodermal explants from, 128–132, *130*, Pl. 6.2
eggs of, 5, 101
gastrula, gastrulation, 101, 102–103, *104*
induced spawning in, 98–100
  injection procedure for, 99–100, Pl. 6.1
lens regeneration in, 142–149, Pl. 6.4
maintenance of, 97–98
metamorphosis of, 107, *108*, 109
tadpoles, 106, *107*, *108*
  hatching of, 106
X-rays, effect on chick embryos, 178

Yolk, of *Aphyosemion*, 86, 87
  axolotls, 112
  chick, 168
  Lebistes, 90
  Xenopus, 101
Yolk plug, of amphibian gastrulae, Axolotl, 114
  Xenopus, 103, *127*
Yolk sac, of chick, 169
  of mouse, cavity of, 207
  of rabbit, 222
    cavity of, 220, 220–222, 223
    stalk of, 225, 226, Pl. 8.2
Yolk sac splanchnopleur, of mammals, 219
  of rabbit, 221, *221*–223, 225, 230, Pl. 8.1, Pl. 8.3
    endoderm of, *222*, 223–226, 229, Pl. 8.3
    localization of immunoglobulin in, 229, Pl. 8.3
    mesenchyme of, 229, Pl. 8.3
    selection of protein by, 230

Zooea, of crab, 75
Zona pellucida, of mouse egg, 199, 206, 208, Pl. 8.2
Zygotene, of mammalian oocyte, 193

Author Index

Adams, E., 118
Albrecht, F. O., 55
Albright, J. F., 158
Allen, E., 192
Amoroso, E. C., 186
Anderson, D. T., 52, 50
Armstrong, P. B., 86
Ashburner, M., 69, 72
Ashby, G. J., 62
Austin, C. R., 206

Baker, J. R., 44
Balinsky, B. I., 200
Barras, R., 54
Bebbington, A., 105, 140
Becker, H. J., 69, 72
Beerman, W., 69
Berendes, H. D., 70
Berg, W. E., 15
Berrill, N. J., 22
Biggelaar, J. A. M., 30
Biggers, J. D., 215
Billett, F. S., 118, 122, 129, 131, 132, 178
Bolton, E., 214
Bonhag, P. F., 51
Boon-Niermeijer, E. K., 30
Boot, L. M., 192
Boveri, T. H., 42
Bowman, P., 179
Brahma, S. K., 129, 132
Brambell, F. W. R., 223
Bretschneider, L. H., 90, 91
Bridges, C. B., 72
Bronson, F. M., 186, 187, 193
Burtin, P., 149
Butt, F. H., 50, 63

Campbell, J. C., 148, 157, 158
Child, H. S., 86

Clark, K. E., 75
Clayton, R. M., 149, 157
Clement, A. C., 30
Clever, U., 69, 70
Cohen, J., 81
Collier, J. R., 30
Collini, R., 178
Colucci, V., 143
Conklin, E. G., 14, 30
Counce, S. J., 50
Crampton, H. E., 29
Cuttler, R. H., 214

Dalcq, A. M., 213
Dagg, S. C., 186, 187
Daniel, J. C., 213
Davies, J., 221, 222, 223
DeBeer, G. R., 12
Demerec, M., 67
Deuchar, E. M., 101
DeWit Duyvene, 90, 91
Dobzhansky, T. H., 69, 72
Dossel, W. E., 8, 12
Drake, J. W., 93
Driesch, H., 14
Dunn, L. C., 69, 72
Duzynskaya, I. G., 52

Eales, H. B., 27
Edwards, R. G., 194, 213
Etkin, W., 160
Evans, A., 91, 92

Faber, J., 101, 104, 105, 107, 108, 143
Fabricius, H., 166
Farr, A. L., 151
Fowler, R. E., 194
Fraenkel, G., 64
Freeman, G., 143, 146, 147, 148
Fretter, V., 31, 35

Frieden, E., 160

Galtsoff, P. S., 12, 76
Gardner, R. L., 214
Giudice, G., 15
Grabar, P., 149
Graham, A., 31, 35
Graham, C. F., 199, 214
Green, J., 75, 77
Gurdon, J., 122

Hale, L. J., 122
Hamburger, V., 12, 112, 115, 116, 117, 121, 174, 177, 180
Hamilton, L., 178
Hamilton, M. L., 174, 177
Harrison, R. G., 112, 115, 116, 117
Harvey, W., 166
Heath, H., 77
Heller, H., 90
Hems, J., 82
Hervey, G. F., 82
Hess, O., 30
Hillman, N., 214
Hinchliffe, J. R., 48
Holtfreter, J., 121, 127
Horstadius, S., 141
Humphrey, R. R., 109
Hunter-Jones, P., 54
Huxley, J. S., 12

Ikeda, A., 157, 159
Imms, A. D. C., 50
Ingram, A. J., 142

Johnson, G. E., 46
Johannsen, O. A., 50, 63
Jones, K. W., 148

Karlson, P., 66, 70
King, T. J., 141
Kostomarova, A. A., 93
Krause, G., 50
Kunz, W., 55

Lillie, F. R., 169
Lowry, O. H., 151

MacBride, E. W., 12, 40, 75
Mahoney, R., 171
Mangold, H., 140
Marsden, H. M., 193
Millar, R. H., 22, 23, 27, 28

Mintz, B., 214, 218, 219
Mossman, H. W., 186, 219
Muller, 42
Mulnard, J., 213
Mystakowska, E. T., 214

Nairn, R. C., 227
Nicholas, J. S., 94
Nieuwkoop, P. D., 101, 104, 105, 107, 108, 143
Nowicka, J., 199
New, D. A. T., 91, 168, 172, 179

O'Brien, B. R. A., 179
Oppenheimer, J. M., 86, 94
Otter, G., 46

Pafenyk, J. W., 85
Palm, J., 194
Pener, M. P., 57
Penners, A., 38, 39
Pugh, D., 179
Purdy, J. M., 213
Purser, G. L., 89

Randall, R. J., 151
Raven, Chr. P., 30
Ray, P. M., 216
Reverberi, G., 12, 15, 36
Romanoff, A. L., 169
Roonwall, M. L., 57
Rosebrough, N. J., 151
Rosenthal, H. L., 89
Rugh, R., 12, 93, 117, 180
Runner, M. N., 194

Sander, K., 50
Saxén, L., 133
Scheidegger, J. J., 154
Schwind, J., 140
Scott, C. M., 61
Scott, W., 216
Seagrove, F., 40
Schaffer, B. M., 160
Scheel, J., 83
Sherman, M. I., 214
Shulov, A., 57
Sinnott, E. W., 69, 72
Slifer, E. H., 57
Smithberg, M., 93
Smyth, J. D., 42
Snell, G. D., 186, 187
Spemann, H., 133, 141

AUTHOR INDEX

Spiegel, M., 125
Spratt, N. T., 179
Steinberg, M., 121
Stephenson, J., 37, 38
Steptoe, P. C., 213
Sterba, G., 83
Stern, M. S., 214
Stevens, L. C., 186

Takata, C., 158
Tarkowski, A. K., 199, 213, 214, 218
Tata, J. R., 160, 162
Taylor, E. R., 44
Thompson, T. E., 105, 140
Toivonen, S., 133
Trinkaus, J. P., 93, 94
Truman, D. E. S., 157
Turner, B. J., 85
Turner, S. C., 142
Tyler, A., 17
Tyler, B. S., 17

Uvarov, B., 57

Van de Lee, S., 192

Vogt, W., 140
Von Euler, U. S., 90

Waddington, C. H., 50
Waggoner, P. W., 149
Weber, R., 160, 162
Welch, P. S., 39
Wessels, W. K., 12
Wheeler, M. R., 67
Whitten, J., 74
Whitten, W. K., 193, 215
Wigglesworth, V. B., 50, 63, 64
Wild, A. E., 223, 224, 229, 230
Wilson, I. B., 214
Wilson, E. B., 30
Wilt, F. H., 12
Witkowska, A., 199
Wolff, F. C., 143, 166
Wourms, J. P., 84
Wroblewska, J., 213

Yajima, H., 52
Yamada, T., 129, 143, 158, 159

Zur Strassen, O., 42
Zwaan, J., 158, 159